THE STORY OF

MINING IN CORNWALL

THE STORY OF MINING IN CORNWALL

A World of Payable Ground

ALLEN BUCKLEY

CORNWALL EDITIONS
FOWEY

This book is dedicated to the miners of Cornwall with whom I have worked and the generations of fine men and women who went before them, who created an industry and a Cornish community all over the world.

CORNWALL EDITIONS LIMITED
8 Langurtho Road
Fowey Cornwall
PL23 1EQ UK

www.cornwalleditions.co.uk
Publisher: Ian Grant

ISBN 1-904880-20-7

Typeset in New Caledonia 11/18 pt
Designed by Maggie Town at One2Six
Editors: Romilly Hambling,
Susannah Marriott, Judy Spours
Illustrations by David Ashby
Maps by ARKA Cartographics
Index by Sue Lightfoot

Frontispiece: Mervyn Randlesome and Micky Roberts on the rearing of a shrink stope above the 315fm level on Dolcoath South Lode, South Crofty mine.

Colour origination by Butler and Tanner, UK
Printed and bound by Butler and Tanner, UK

Papers used by Cornwall Editions are natural, recyclable products made from wood grown in sustainable forests; the manufacturing processes conform to the environmental regulations of the country of origin.

CONTENTS

FOREWORD

This volume marks a significant step forward in the study of Cornish mining history. For the first time since the foundations of the subject were laid in the mid-20th century, Allen Buckley brings together a greatly expanded literature to produce a new and extended overview. In doing this he takes a number of new initiatives. This is, for example, the first sustained attempt to embrace the entirety of the subject's history, from the most ancient to the modern period: the first to examine all aspects of the industry in one integrated volume, embracing quarrying and the minor metallic minerals alongside the better known copper and tin; and the first to try systematically to set the industry within the wider political and social context of the county and country.

The book is not simply a synthesis of the work of others. It includes much that is new from Allen's own original and otherwise unpublished research, which extends and challenges many established interpretations. His comments on the often changing roles of miners working under the Cornish employment system provide a particularly good example. Gleaned from a careful reading of a wide range of original sources, they challenge the established interpretation of the subject and will require a significant revision, not just in Cornwall but in mining districts across the world where that system was so commonly adopted.

Allen Buckley brings an authority to mining history shared by few others. He has been an active contributor to the subject over four decades, authoring numerous books and articles of his own and, through his long editorship of the *Journal of the Trevithick Society*, helping to shape and focus its continuing development. He is an academically trained historian, with considerable experience of making his knowledge available to a wide audience. Perhaps most important of all, however, he has had extensive experience of mining first hand – he is a writer who really does know what he is talking about!

Within the broader context of Cornish studies, this volume will complement recent work on the remarkable outflow of population to other parts of Britain and abroad. By providing a clearer insight into the causes of the decline of the local industry, and the advanced technical skills possessed by Cornish miners, it further advances our understanding of the complex mix of factors that both forced and facilitated those movements. Yet again we are reminded that, even in the early industrial age, Cornwall and the Cornish existed in a global economy and the forces that shaped its economic and social history were as much generated outside of the county as within it. This new road map of where we are in the development of the subject will no doubt stimulate a new generation of researchers and guarantee a central position for Cornish mining, not only in the industrial history of Britain, but of mining communities everywhere.

Roger Burt
Emeritus Professor of Mining History
University of Exeter

INTRODUCTION

OPPOSITE
A magnificent view of the enormous gunnis on Great Lode at the 255fm level of Wheal Agar, part of East Pool mine. The stope varies from 12-15 m (40-50 ft) wide and is some 30.5 m (100 ft) in height. It was mined entirely by hand labour and blasted out with gunpowder nearly a century ago.

FEW INDUSTRIES CAN RIVAL the antiquity of Cornish mining, and still fewer have excited such fascination among economic and social historians. The origins of the Cornish tin industry are lost in the mists of prehistory, and much of the romance which has attached itself to the story has to do with the mystery many feel when contemplating that 'other world' deep beneath the beautiful Cornish landscape. The ancient land of Cornwall, with its legends and folklore, encourages belief in such things as the visits to Cornwall of the Phoenicians, of Joseph of Arimathea and the boy Jesus. Myths abound in such a place and in such an atmosphere. The truth is more prosaic but no less exciting, and when set in the context of the development of western civilisation and the multitude of inventions and innovations that have marked its progress we learn to appreciate the significant role which the Cornish metalliferous mining industry has played in that long and fascinating story.

Making metals from the ores found in the earth was one of the earliest major steps taken by man to cope with and control his often hostile environment. Probably the oldest extant reference to the production and use of metals for man's benefit is to be found in the first book of the Bible, *Genesis*, chapter four. There it talks of Tubal-Cain's skill as a metal worker. Biblical references to the use of various metals, such as copper, iron and tin, become more frequent as the history of those ancient Hebrew peoples is related. Mining is described in some detail in the *Book of Job*, and it is a fascinating fact that archaeological finds from ancient tin sites in Cornwall indicate that the tin industry was already in existence at the time when that description was written down.

Several authors have contributed to the growth of interest in Cornish mining history, and for nearly five hundred years they have described or commented on aspects of it. John Leland, Thomas Beare, Richard Carew and John Norden in the 16th century, the contributors to the *Philosophical Transactions of the Royal Society* in the 17th century, and Thomas Tonkin, William Borlase and William Pryce in the 18th century, were all fascinated by this unique story. During the 19th century interest really blossomed, as W.J. Henwood, George Henwood, J.W. Colenso, J. Carne, Thomas Spargo, J.H. Collins and a host of others wrote books, articles and learned treatises on the subject. However, it was in the early years of the 20th century that Cornish mining history, as we now know it, really took off. In the early years of that century G.R. Lewis wrote *The Stannaries*, and a whole new world of research into the medieval Cornish tin industry opened up. In 1927 A.K. Hamilton Jenkin wrote *The Cornish Miner*, and the subject suddenly took on a new appearance. Here was

a book that was not just about production, ore reserves, techniques, managers and mineral owners but one which also considered the miners and their lives at work and at home. Jenkin examined the whole story as a historical subject and not just as a source of information for potential investors or moral improvers.

In the 1960s, after Jenkin's 16 volumes of *Mines and Miners of Cornwall* appeared, a series of books was published by Bradford Barton of Truro. These built on the work of Jenkin and examined, sometimes in detail, many aspects of Cornish mining history. *The Cornish Beam Engine, A History of Tin Mining and Smelting in Cornwall, A History of Copper Mining in Cornwall and Devon* and two volumes of essays on aspects of Cornish mining history were published. Barton also published books by such serious students as Bryan Earl and Arthur Raistrick and reissued several titles long out of print, like William Pryce's *Mineralogia Cornubiensis* and G.R. Lewis' *The Stannaries*. Since then others have made valuable contributions to the wonderful story of Cornwall's metalliferous mines and the men who worked them.

This book builds on the efforts of these authors and many others who have spent long hours over decades researching particular areas of interest or expertise and recording their conclusions for other historians. Works such as Roger Penhallurick's fine *Tin in Antiquity* and John Hatcher's unrivalled *English Tin Production and Trade Before 1550* have been been used extensively in preparing this book, and I encourage readers who are interested in furthering their knowledge of these subjects to turn to them. When I look back at what has already been accomplished I feel I stand on the shoulders of giants. I hope my efforts to use their knowledge and research to set out the story of Cornwall's mines and miners will better the reader's appreciation of this most fascinating subject.

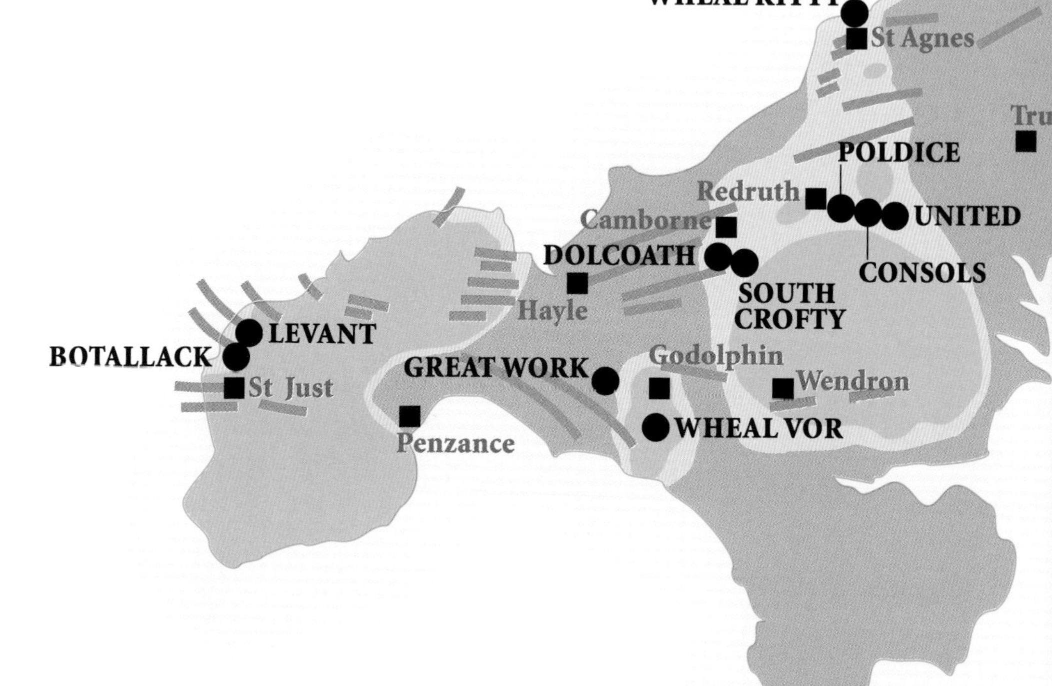

Cornish mining industries have their source in the metamorphic rock that surrounds each major outcrop of Cornwall's spine of granite. The map shows the major mines that have exploited the many metal-bearing lodes revealed diagrammatically here, from Botallack and Levant in the west to Devon Great Consols in the east. The whole of Cornwall has been involved in the extraction of tin, copper, lead, silver, zinc, iron, slate and china clay. The map demonstrates just how significant the mining industry was for the whole of Cornish society over many centuries.

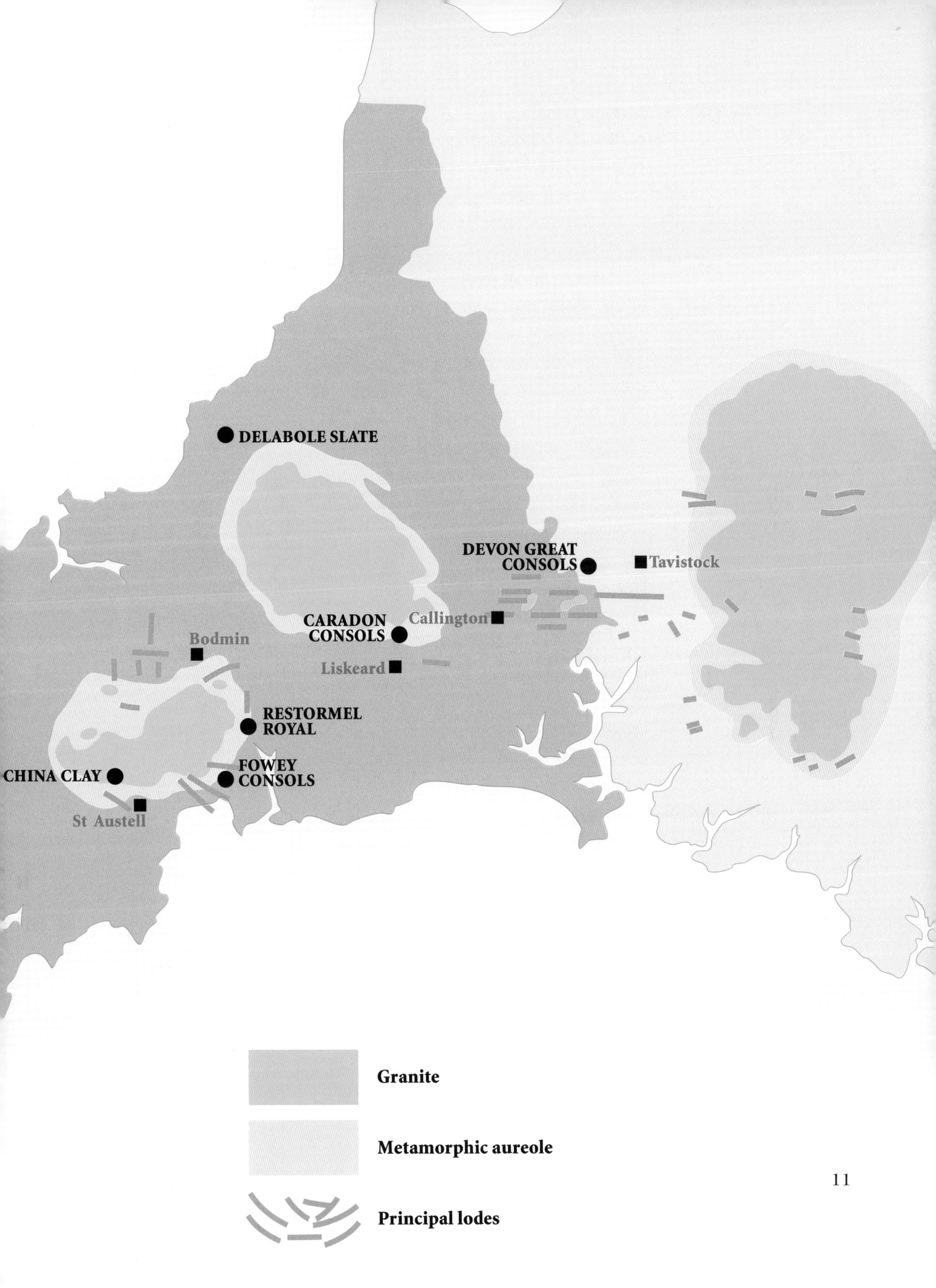
DELABOLE SLATE
DEVON GREAT
CONSOLS
Tavistock
CARADON
CONSOLS
Callington
Bodmin
Liskeard
RESTORMEL
ROYAL
CHINA CLAY
FOWEY
CONSOLS
St Austell
Granite
Metamorphic aureole
Principal lodes

CHAPTER ONE

THE ANCIENT TIN INDUSTRY

Prehistory to the Norman Conquest

OPPOSITE
This fine example of tin ore contains beautiful cassiterite crystals with quartz. The piece was found at Carn Brea Mine, Illogan Parish, and can be seen at Truro Museum.

PRECISELY WHEN MANKIND learned the art of processing crushed ore-bearing material into a concentrate and smelting it into metal is not known. The evidence suggests that men were developing such skills several thousand years ago, and that the working of metals into household utensils, tools and weapons swiftly followed. Copper was discovered and exploited early on and its use spread quickly throughout the civilised world. The ubiquitous copper-bearing lodes were easily discerned by the intelligent observer, and very soon all manner of useful objects were in circulation. However, copper is a relatively soft material and not ideal for weapons, and when we learned to mix it with tin a harder and more durable metal was produced. This alloy was bronze, and its relative hardness made it ideal for tools like axes, knives and chisels, and, more importantly, for weapons, such as swords and spears.[1]

Tin-bearing lodes or veins are not common and there were few sources for the metal in the ancient world. Some tin was mined in Asia Minor more than 3000 years ago, and from the 1st century AD it was worked in the Roman mines of Spain, but for 2000 years before the Christian era the south-west corner of the British Isles was one of the main sources of tin for the civilisations around the Mediterranean Sea.[2]

The earliest evidences of a Cornish tin industry are from archaeological finds on old tin-working sites. The practice of reworking long-abandoned stream works has resulted in the discovery of artifacts which have provided proof of tin working in prehistoric times. These finds have covered the whole timescale from the Early Bronze Age through the Romano-British period to the time when the Saxons dominated Cornwall. Finds from the Early Bronze Age (c. 2100–1500 BC) have been tantalisingly few, but they are sufficient to provide proof that tin was produced from the alluvial (formed by the action of water) deposits of Cornwall 4000 years ago.

Bronze Age finds

The four Early Bronze Age sites which have provided evidence of tin working are at Levalsa Meor (between Pentewan and St Austell), the Carnon Valley, St Erth, and Caerloggas Down near St Austell. The artefacts found consist of a jet slider or belt

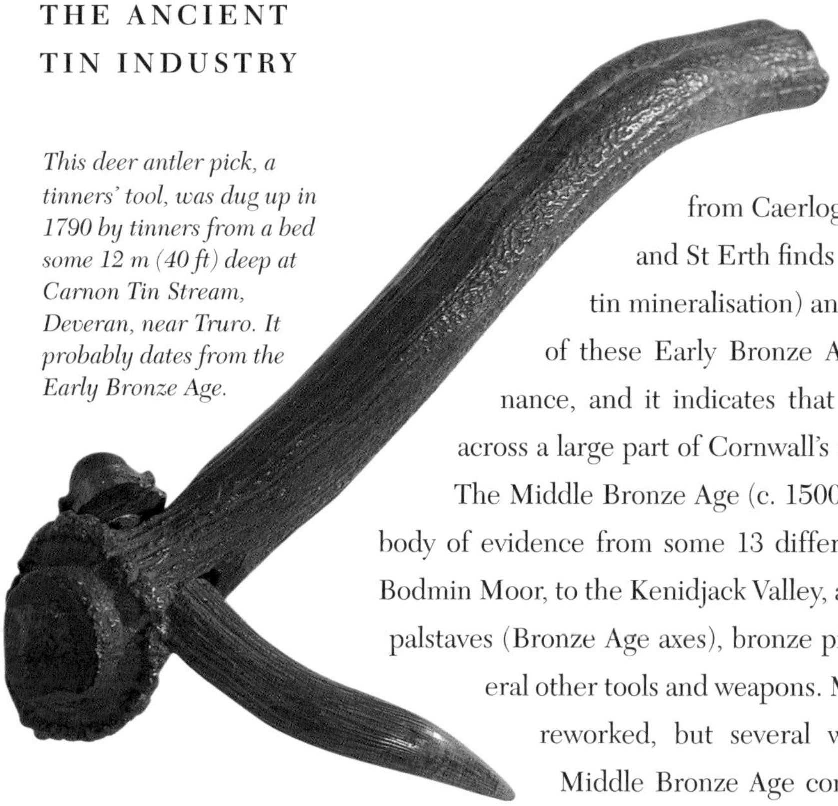

This deer antler pick, a tinners' tool, was dug up in 1790 by tinners from a bed some 12 m (40 ft) deep at Carnon Tin Stream, Deveran, near Truro. It probably dates from the Early Bronze Age.

fastener from Levalsa Meor, an axe and two antler picks from the Carnon Valley, a stone axehead from the Hayle River above St Erth, and some cassiterite and tin slag from Caerloggas Down. The Levalsa Meor, Carnon Valley and St Erth finds were all discovered on tin ground (areas with tin mineralisation) and well below the surface. The context of each of these Early Bronze Age objects has established the find's provenance, and it indicates that during that remote period tin was worked across a large part of Cornwall's stanniferous (tin-bearing) districts.[3]

The Middle Bronze Age (c. 1500–800 BC) has supplied us with a much larger body of evidence from some 13 different locations spread between Bolventor, on Bodmin Moor, to the Kenidjack Valley, at St Just in Penwith. The finds range through palstaves (Bronze Age axes), bronze pins, socketed chisels, spears, rapiers and several other tools and weapons. Most were found on tin ground that was being reworked, but several were discovered on archaeological sites in Middle Bronze Age contexts. This increased occurrence and wider distribution suggests that the Cornish tin industry was well established throughout Cornwall more than a thousand years before the birth of Christ.[4]

The Late Bronze Age (c. 800–500 BC) has left less archaeological evidence of tin activity than the previous periods, but absence of evidence is not evidence of absence and, undoubtedly, production of tin continued at that time much as it had in previous centuries. In March 1792 two bronze cauldrons were discovered at Broadwater, Luxulyan, on the headwaters of the Par River. They lay at a depth of 8.5 m (28 ft) and were in good condition. Penhallurick believed them to have been of local manufacture, but it is of interest that two eminent archaeologists, C.F.C. Hawkes and M.A. Smith, believed the cauldrons were of the type that came from Massalia (Marseilles) and were dated at about 600 BC.[5]

A tinners' oak shovel, which was found by tinners on Boscarne Tin Stream in the 19th century. It is thought to date from between AD 710 and 910. It is now on exhibition at Truro Museum.

This half of a rapier mould was ploughed up at Bodwen Farm, Lanlivery in 1971. The molten metal was poured in through the depression at the pointed end. Diagonal grooves allowed the gasses to escape. It dates from between 1450 and 1200 BC and is evidence of bronze weapon-making locally, and possibly of prehistoric mining of Cornish copper.

Iron Age finds

For the Iron Age (c. 500 BC–43 AD) we have information from two distinct sources: archaeological finds and written records. The archaeological evidence comes from a variety of sites, three of which – Levalsa Meor, Trenowth on the Fal River and Red Moor in Lanlivery – were tin-working sites, and three were from archaeological digs unrelated to tin works. The evidence suggests that all the finds can be dated to the Iron Age. The Levalsa Meor find, an ash tankard with bronze holding strips and handle, was found 8.5 m (30 ft) below the surface. Iron Age tin spindle-whorls were found at Bussow on the Land's End peninsula and at St Columb Minor.[6]

The written evidence from the Iron Age is of special interest not least because of the link, referred to above, made by Hawkes and Smith to the Broadwater cauldrons and Massalia. Towards the end of the 4th century BC (c. 325–306 BC) a Greek explorer-geographer known as Pytheas of Massalia travelled to the British Isles and is said to have reached as far north as Iceland (Thule). Pytheas circumnavigated Britain, describing the northern tip of Scotland as 'Orka' (opposite the Orkney Isles), the south-eastern edge as 'Kantion' (Kent) and the south-western corner as 'Belerium' (Land's End peninsula). Although Pytheas' writings are no longer extant, it is generally accepted that the account by Diodorus Siculus in the 1st century BC was based on his writings. The translation by C.H. Oldfather (1939) reads:[7]

> The inhabitants of Britain who dwell about the promontory known as Belerion are especially hospitable to strangers and have adopted a civilised manner of life because of their intercourse with merchants of other peoples. They it is who work the tin, treating the bed which bears it in an ingenious manner. This bed, being like rock, contains earthy seams and in them the workers quarry the ore, which they then melt down and cleanse of its impurities. Then they work the tin into pieces the size of knucklebones and convey it to an island which lies off Britain and is called Ictis; for at the time of ebb-tide the space between this island and the mainland becomes dry and they can take the tin in large quantities over to the island on their wagons. (And a peculiar thing happens in the case of the neighbouring islands which lie between Europe and Britain, for at flood-

> tide the passages between them and the mainland run full and they have the appearance of islands, but at the ebb-tide the sea receded and leaves dry a large space, and at that time they look like peninsulas.) On the island of Ictis the merchants purchase the tin of the natives and carry it over from there across the Straits of Galatia or Gaul; and finally, making their way on foot through Gaul for some thirty days, they bring their wares on horseback to the mouth of the river Rhône.

This detailed description is extremely informative about the Cornish tin trade of the 4th century BC, and although various passages need elucidation, the main points are clear enough. Argument has continued for decades over the meaning of 'This bed, being like rock, contains earthy seams and in them the workers quarry the ore'. Some have seen it as describing lode working, for many tin lodes are, by comparison with the host rock, soft and earthy. Most, however, believe it refers to working the alluvial deposits of tin. The reference to working the tin into 'pieces the size of knucklebones' is harder to understand. Other translators interpret the word *rhuthmos* as referring to the shape rather than the size of the pieces of tin, and this makes more sense than Oldfather's translation.

What is certain is that in the 4th century BC the inhabitants of Cornwall were operating a sophisticated tin-producing industry and supplying a well-established export network. They were locating the tin-containing deposits, organising the labour involved in removing overburden and extracting the tin-bearing material, arranging the dressing of the black tin (cassiterite concentrate), and then smelting it to produce white tin metal. They had markets organised, transport arranged and customers several hundred miles away to buy it. A.C. Thomas identified Corbilo, at the mouth of the River Loire, as the port of entry into Gaul. From thence the merchants carried the tin on horseback through the Loire Valley to the headwaters of the River Rhône, whence the route went south to Massalia. This journey took a month. To meet the needs of merchants who travelled many hundred miles to purchase the tin

*The St Mawes ingot, dredged up by fishermen in 1812 from Carrick Roads, Falmouth, and taken to Calenick smelting house in Truro for analysis. It weighs 72 kg (158 lb). Its peculiar shape causes association with the knuckle-bone-shaped ingots (*astragulus *in Greek) mentioned by Diodorus Siculus.*

metal, the industry must have been in a settled, fairly permanent and reliable condition. Dressing the tin-bearing sand to produce black tin concentrate was a highly skilled operation. The accurate construction of furnaces and the ability to raise the temperature to more than 1,000°C to produce pure tin metal were not easily acquired skills.[8]

THE PHOENICIAN MYTH

By the 19th century the idea that Phoenician merchants traded directly with west Cornwall in the centuries before Christ was well established. Not only that, but it was widely accepted among respected historians and other educated people that many Cornish words and practices had been introduced by these ancient people. The tin used to make the bronze implements in Solomon's temple was also believed to have come from Cornwall, and, to add colour to the story, it became a commonplace that Joseph of Arimathea, the supposed uncle of Jesus Christ, travelled to Cornwall on Phoenician trading vessels to purchase tin. It took a small step to have Joseph bringing his nephew to Cornwall, undoubtedly to teach the young man the tin trade. With these ideas firmly fixed in the minds of the great and good of Cornwall, it came as something of a shock to be told, at a meeting of serious historians, that the whole story had no basis in fact. The *Journal of the Royal Institution of Cornwall* reported the announcement in May 1863 and informed its readership that the news was greeted with 'astonishment'.

In his book *Tin in Antiquity* Roger Penhallurick offered convincing proof that, although there is an abundance of evidence for the ancient tin trade between Cornwall and the western Mediterranean, there is none for direct trade between Cornwall and the Phoenician ports of Tyre and Sidon in the centuries before Christ. Further to that, Penhallurick showed that the tin trade to Massalia (Marseilles) from Cornwall was in the hands of Greek traders, not Phoenicians, and the writings of Diodorus Siculus (first century BC), based on accounts by Pytheas and Timaeus (from the third and fourth centuries BC) support this. Penhallurick further shows that the so-called evidences of a Phoenician involvement with Cornwall have different and much more reasonable explanations. Despite the fact that well over a century has passed since the Phoenician myth was first challenged and no serious historian now accepts it as historical, belief in the story persists and we still occasionally hear fairly well-educated people maintain that the ancient Phoenicians traded with Cornwall before the birth of Christ.

In support of the supposition of an early and continuing tin trade between Cornwall and Massalia, coins from that city have been found in the tin-bearing districts of both Brittany and Cornwall. Two coins found at Plouguerneau, Finistère, originated in Massalia in the 2nd century BC, and a hoard of 45 silver coins was discovered in 1909 at Paul, just west of Penzance. These silver coins appear to be copies of those made in Massalia, were probably struck in northern Italy, and have been dated to the third century BC. Thomas has proved the existence of trade connections between Cornwall and Brittany at that time. The fact that the export trade could already have been established for 300 years before the time of Pytheas – if the bronze cauldrons indeed came from Massalia in about 600 BC – indicates that the Cornish tin trade was a very old one and not a localised and haphazard affair.[9]

A contemporary of Pytheas, Timaeus of Sicily (c. 352–256 BC), was quoted by the Roman writer and historian Pliny (AD 23–79) in his *Natural History*, and although his description of British tin is vague, it confirms that it certainly was of consequence in the third and fourth centuries BC. Pliny wrote: 'Timaeus the historiographer saith, that farther within at six days' sailing from Britannia, is the island Mictis, in which white lead is produced, and that the Britanni sail thither in wicker vessels sewed round with leather.' It seems that the 'white lead' was tin metal, as the Romans do not appear to have had a separate word for tin, the word *stannum*, apparently, being borrowed from the natives of Cornwall, who used the word *sten* for tin. Many believe that the 'Mictis' of Timaeus is the same island as Pytheas' 'Ictis', and the most popular identification is with St Michael's Mount, adjacent to some of the most productive ancient tin districts.[10]

Finds of the Romano-British period

Archaeological finds from the Romano-British period (AD 43–430) abound. Artifacts have been discovered on at least eight abandoned tin workings from Land's End to Bodmin Moor, and once again the finds are from sites as diverse as the high central moorland, the shallow river estuaries of the south coast and the steeper valleys of the north coast at St Columb Minor. They include a variety of artifacts, such as tin ingots with handles, a pewter bowl, three tin bowls, rings, brooches and a number of Roman coins. Archaeological digs on sites unconnected with tin working have thrown up a whole array of objects that support the case for widespread tin working in Cornwall during the Romano-British period. These finds include tin and pewter jugs, a tin ingot and Roman coins from Marazion Marsh.[11]

Supporting the archaeological evidence is the testimony of Roman writers. These not only refer back to the earlier historical accounts of Pytheas and Timaeus but argue over the significance and even existence of economically viable reserves of British tin as a reason for invading and colonising the British Isles. While Caesar

This late 19th-century picture of St Michael's Mount shows it with the tide out. The ancients referred to the isle of Ictis as off the coast where tin was taken for export to Gaul, and many believe that St Michael's Mount is the ancient Ictis.

asserted that 'Tin is found in the inland parts' of southern Britain, Cicero denied the existence of any valuable metals in Britain. Strabo's list of commodities then exported from Britain pointedly left out both lead and tin. Cicero's and Strabo's arguments appear to have been influenced as much by politics as by the facts. The accepted view of the Cornish tin trade in Roman times is that it was severely disrupted by Caesar's invasion of Brittany and destruction of the Armorican naval and trading fleets in 57 BC. Cornish tin exports were further curtailed during the first century AD as the Romans began to rely on supplies from the expanding Spanish tin mines. Some two centuries later, in the third century AD, there was a rapid expansion of Cornish tin production as domestic demand in Britain increased due to a

THE ANCIENT TIN INDUSTRY

Chun Castle, West Penwith, is a 2nd- or 3rd-century BC fort that contains evidence of tin smelting, although some archaeologists believe that the furnace there is likely to be post-Roman. Chun Castle lies close to several ancient tin-working sites.

Chysauster Courtyard Village, near Penzance, is an Iron Age settlement on the side of a valley, which was worked for alluvial tin in prehistoric times. Smelted tin and pottery with tin glaze have been found in huts here.

The Carnanton tin ingot was discovered in 1819 and found to contain a Roman imperial stamp. It has been dated to the 3rd or 4th century AD, and it furnishes proof of Roman control of the Cornish tin trade. The ingot can be seen at Truro Museum.

greater use of pewter for eating and drinking vessels among the wealthy and middle classes. This coincided with a decline in production from the Spanish mines, which left Rome's supply vulnerable. The discovery at Carnanton of a 3rd/4th-century AD Roman tin ingot bearing an imperial stamp is a sure indication that the British tin trade was controlled at that time by the Roman authorities. Roman forts, milestones and other memorials from the 3rd and 4th centuries support this, as does the Roman villa at Magor, on the hillside above the Red River, a rich source of alluvial tin since prehistoric times.[12]

Dark Age finds

The long period that followed the withdrawal of the Romans from Britain is usually called the Dark Ages (AD 430–1066). This is because historians of earlier centuries believed that few written records had survived from that time, and, although this is certainly not the case, it is true that the country entered what many consider a dark period in the island's history. Archaeological evidence of tin working from that time is relatively limited, leading some to the conclusion that little was going on. However, among the finds on old tin-working sites are a Saxon silver coin hoard, found near Pentewan, a ring and brooch from Paramoor Valley, a brooch from Lanivet and an oak shovel found at Boscarne, near Bodmin. Evidence of tin smelting during this period has also been found at Chun Castle, on Land's End peninsula, and in the form of tin ingots from Praa Sands. Evidence of Dark Age tin smelting on Dartmoor and tenth-century tin working at Lanlivery has also been found. These finds provide no indication of the extent of the industry during those centuries, but they confirm that tin working continued, albeit on a smaller scale than hitherto.[13]

A manuscript that supports the continuing existence of Cornish tin production during the seventh century AD is known as the *Acta Sanctorum*, and concerns St John the Almsgiver. The document tells of a shipload of tin from this country which was to be taken to Egypt to be traded for corn. The account contains miracles and other historically questionable material, but it does confirm that Cornwall was regarded as a source of tin at the time and that it had trade links with the eastern end of the Mediterranean. Hatcher informs us that tin was exported from Cornwall and Devon to Russia in the ninth century, and the transalpine trade in this tin, conducted by the Lombards, was flourishing in the tenth century. Thomas Beare, writing in 1586, expressed his belief that Saxons had been involved in the tin business. 'It appeereth by working of our tinners in Cornwall that the Saxons being heathen people (when they inhabited our country) were skillfull workers of and sercers for black tyn, which in those auncient dayes wrought not with spades & working tooles made of iron as we have now in our tyme but all of the hart of oake, they as they got their tyn and their blowing houses and places hard by their works and so made it white, for prooff where of diverse workers of our tyme have fownd their shovells spades & mattox made all of oke & holme in divers and sundry places.' Beare may have been correct, but if the Saxons were involved tin production they have unfortunately left little evidence of it. Richard Carew, writing in 1602, appears inadvertently to support Beare's belief, for although he suggests that it was the Jews who first worked many of the tin works where ancient tools were found, he tells us that such workings were called *Attal Sarazin* by the old Cornish tinners. However, Carew was mistaken in his translation of the Cornish, for the words appear to mean 'Saxon offcast', not 'Jewish offcast' (offcast was the material left after smelting).[14]

Although we know little about the organisation of the tin industry in Saxon times, we do find some hints as to the rights and privileges enjoyed by the tinners of those days. For example, when William de Wrotham, on being given jurisdiction over the stannaries by Hubert, justiciar of Richard I, in 1198, reorganised the whole tin industry of Cornwall and Devon, he introduced new taxes, more rigorous controls and new weights and measures. But he also confirmed that the tinners already possessed various customary rights and privileges. He recognised that the ancient tin industry had various accepted divisions of labour, such as diggers, smelters, ore buyers and tin metal dealers, each of which had its own customs. These restrictions, demarcations, liberties and privileges are indicative of a long-established industrial organisation. And, only three years later, the first stannary charter issued to the tinners of Cornwall and Devon by King John in 1201 referred to these ancient liberties when it described the freedom enjoyed by the tinners to dig tin and the turves with which to smelt it and to divert water to operate their workings. Significantly, the Charter continued with phrases such as 'as they have been accustomed', and 'as by ancient custom they have used'. Thus, King John was merely reaffirming rights and liberties the tinners had enjoyed since time out of mind – in fact reaching right back into Saxon times. The lands of bishops, abbots and earls were included in the areas that the tinners could work freely. Clearly, John's government understood the value to the crown of this ancient industry and sought to encourage output by confirming the tinners' privileges.[15]

Techniques and tools

Undoubtedly, in prehistoric Britain most tin came from alluvial and eluvial (refers to material in a soil horizon) workings, although the tinners would not have ignored the rich outcrops of lodes when they were found. Understanding the actual methods used in the production of tin in antiquity is not easy. A wide variety of theories are offered, based on several pieces of evidence from a number of sources. Some believe that the shallower deposits of alluvium, higher up the rivers and streams of Cornwall and Devon, would have been the first to be exploited. Evidence of such early tin workings on Dartmoor and Foweymore (Bodmin Moor) suggest that this might have been the case, and logic supports the idea. However, numerous finds from the Early and Middle Bronze Ages inform us that operations exploiting the deeper seams of tin-bearing material were carried out close to the coast on the wide, flat beds of estuarine alluvium. The location and nature of the alluvial deposits would have determined the methods chosen to obtain the tin-bearing material. Where the deposits were shallow and the valley fairly narrow, it would have been practical to remove the overburden and lay bare the tin ground. This could then be removed systematically for washing, using primitive buddles of a strip type. Where the overburden was thick and the tin ground deep, a totally different method was sometimes employed.

In the middle of the 19th century at Wheal Virgin, more than a mile above Pentewan, tinners uncovered an ancient timbered shaft that went from 3 m (10 ft) below the surface to the tin ground another four to six metres below. The original shaft had been sunk to a depth of 4.5 m (15 ft) and had an oak frame which was fixed together with mortices and tenons. The shaft was described as square in plan, but no other dimensions were recorded. Hurdles of oak twigs had been used to fill the gaps between the timbers, and clay was apparently used to seal the shaft from penetration by water. The shaft appears to have been sunk by Bronze Age tinners and is similar to even older shaft constructions that have been found and dated in Holland. The existence of such a shaft on ancient tin workings raises questions as to its use. It clearly gave access to the tin ground four to six metres below the surface, but once there the tinners could have proceeded in one or more ways. They could merely have exploited the ground at the bottom of the shaft. This is unlikely as it would not have justified the relatively time-consuming and expensive construction of the shaft. Or the shaft could have been used to remove all the tin ground within easy and safe distance from the bottom. In this case the methods that were used would probably have been similar to the more modern 'room and pillar' method of ore extraction. Tunnels would be driven from the bottom of the shaft and pillars (columns of soil or rock) left to support the overburden. At some later point the pillars would be removed as the tinners worked backwards to the shaft base. Occasionally, timber might have been needed to support weak roofs, but mostly this would not have been

This coffin work, or openwork, at Wheal Coates, St Agnes, is a typical example of a tin lode outcrop working. The earliest workings at Wheal Coates are ancient, possibly prehistoric, and although this coffin cannot be accurately dated, it is typical in most respects to this type of lode exploitation.

necessary. As only one such ancient shaft has been discovered (or at least it has been recorded), it may be that the method was rarely used, but it certainly would have been an effective and economical system to employ under most conditions where the tin ground was wide and deep. This possibility that this system was used to work the deep alluvial tin ground is supported by the translation of Diodorus by J.D. Muhly, which presents the crucial passage: 'The bed is of rock, but contains earthy interstices, along which they cut a gallery.'[16]

Where tin lodes were discovered in cliffs or on hillsides, usually by following the 'shoad' stones (pieces of mineralised ore lying at surface) back to the outcrop or by digging 'costean' (exploration) pits to locate the back of the lode, the tinners would have worked them from the surface by cutting a trench, called a coffin or beam work. These openworks would be used to reach the lode for as long as it was practical, after which it would have been necessary to resort to shaft and level mining. With alluvial tin so readily and relatively cheaply available for exploitation, it is not thought likely that the ancient British tinners pursued this method to any great extent, although they would undoubtedly have known of its widespread use abroad from travellers, including the merchants who came to purchase the tin. Mining underground would simply not have been an attractive prospect because of the technically daunting and expensive problems of lighting, ventilation and water removal at a time when there was still plenty of tin available at or near the surface.

The tools used in the alluvial workings were made from a variety of materials. Antler ('hartshorn') picks, oak and holme picks, wooden shovels and bronze chisels have all been found in ancient tin streams. Undoubtedly, the ubiquitous iron poll picks were also eventually used, as they have been continuously in mining and streaming until the present day. When it was abandoned on tin works in wet valley bottoms, iron would tend to disappear through oxidation, so little evidence of their use remains.[17]

Once the tin ground was extracted, the methods of separation and dressing to produce a black tin (i.e. cassiterite) concentrate would have been similar, in some respects, to more recent techniques. Flushing with water remains the best way to wash away the lighter gangue material from the denser and heavier cassiterite. First, water would be used to wash away the mud and lighter sand from the tin-bearing pebbles and sand. The pebbles and small rocks were then crushed by hand using crude tools, or ground with small crazing stones, and the residue carried to trench-like buddles where further waste was removed. Repeated washing, with the water flow controlled to the correct velocity and volume, brought the grade up to a point where the dressed ore could be sent for smelting. This black tin concentrate was probably produced by small groups of tinners working on their own account, much as they did throughout the medieval period.

The smelting operation itself required a different level of skill and perhaps lay in the hands of an elite, who may have been locally or even regionally important. It is possible too that at this point foreign merchants had some control of the tin. The archaeological evidence we have that tin was smelted locally consists of the remains of small bowl furnaces, which were essentially a small pit dug in the ground, where the molten tin went after smelting, and a flimsy structure of stone and clay that formed the furnace itself, over a small pit dug in the ground, where the molten tin ran after smelting. The black tin concentrate was mixed with charcoal and placed in the furnace, where bellows were used to raise the temperature to over 1,000°C (832 °F). It seems that in some cases the prevailing wind was directed through a tunnel into the furnace to increase the temperature. The ingots produced by this method tended to be plano-convex in shape – a bit like a pasty – and several can be seen in Cornish museums. The smelting technique was used well into the medieval period.[18]

Customs, practices and celebrations of the tinners

The Dark Ages saw several changes in the lives of the tinners. In the fifth or sixth century AD the so-called Celtic saints are believed to have arrived from Wales, Ireland and Brittany, and they brought with them a new religion – or at least a new version of an old religion. The lives of these saints, some of which, like *Bewnans Meriasek* and *Bewnans Ke*, have come down to us, suggest that the early Christianity that had been brought to Britain in the second and third centuries was still, to some extent, practised when these saints arrived. Wonderful stories about the 'Celtic saints', many of which have survived until our day, circulated among Britons

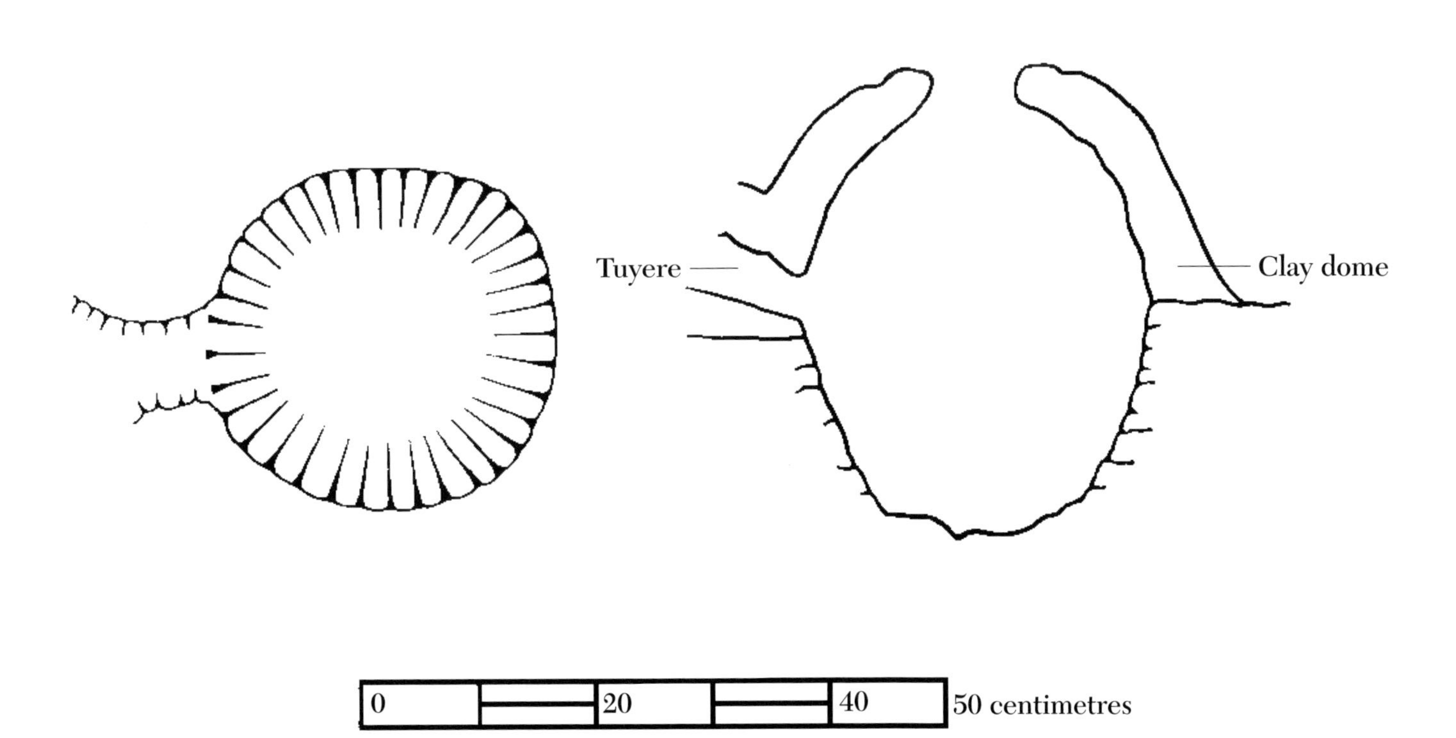

This diagram, from R. F. Tylecote's original drawing, of a bowl furnace reveals that ancient tin smelting furnaces consisted of a small pit in the ground to receive the molten tin, with a crude superstructure of stone and clay forming a dome over it. Cassiterite and charcoal were placed in the furnace and fired. Air was blown in to increase the temperature.

YOW WHIDDEN

Yow Whidden was one of the great tinners' festivals. In Cornish it means 'White Thursday', and it was celebrated in west Cornwall on the Thursday before Christmas. It seems that in east Cornwall it coincided with Picrous Day, although some believe that that fell on the second Thursday before Christmas. *Whidden* (*whydn, gwyn, widden, whidn*, etc.) literally means 'white' in Cornish, but it can carry the connotation of 'happy' or 'good'. As *Yow* means 'Thursday', the meaning of the festival would appear to be 'Happy (or Good) Thursday', and as Christmas follows quickly after it, this might be what is referred to. Sometimes the festival was called 'Chy Whidden', or 'White House', which could be a reference to the tin-whitening house or blowing-house. In support of this is the connection with the discovery of tin smelting traditionally attributed to St Piran, together with his associates St Picrous and St Chywhidden. It seems likely that the two festivals developed separately in east and west Cornwall, celebrating the same half-forgotten events and heroes at the same time of the year. As late as the 1820s there are records of tinners taking the day off work and celebrating – with much drinking, eating, merrymaking and singing – the traditional day when tin working first began.

Other festivals were celebrated throughout the year. Old Christmas Day was remembered at Epiphany, on 6 January, and Paul's Tide or Paul's Pitcher Day was remembered the day before the Feast of St Paul (24 January). St Piran's Day, on 5 March or the nearest Friday, was also celebrated as Friday in Lide (the first Friday of March). Midsummer Day was, and still is, remembered with a great bonfire.

While all of the festivals were celebrated boisterously, some had their own very special features. On Paul's Pitcher Day, the tinners stood an old pitcher on a bank and pelted it with stones until it was completely destroyed. They then headed for the nearest pub where a new pitcher was obtained, filled with beer and emptied several times during the course of the day. These proceedings seem to be related to the tinners' abstinence from alcohol while at work on their tin bounds (areas in which they had acquired the right to mine – see Chapter Two). The new pitcher was used during the following year to carry drinking water to their place of work.

On the Friday in Lide, or St Piran's Day, a young lad was sent to sleep in the burrow (pile of mine waste) within the tin bounds. The length of time he slept determined the normal length of the tinners' post-croust (croust was the mid-shift tea break) snooze for the following year. Needless to say, the early March weather would not have encouraged a long sleep![19]

of the time. Some of these legends and myths concern the tinners, whose customary practices and annual celebrations also appear to have originated at this time. Thus, traditionally, St Piran is supposed to have been the man who taught the Cornish to smelt tin. The tinners' celebrations were held on or close to important days in the Christian calendar, or they marked changes in their annual routine. The religious beliefs and superstitions of the tinners, most of whom lived their working lives in remote places on high moorlands or secluded valleys, became intermixed with the observances and practices of the relatively isolated Christian communities of Cornwall, but they also developed their own unique customs, traditions and celebrations. The tinners' way of life and daily routine were peculiar to them and marked them out as separate and different from the population at large. For example, outsiders were called 'foreigners', and even the medieval charters granted to the tinners referred to them in this way.[19]

Tinners' traditions date from a remote period before records began, and although most were in place by the Dark Ages, they probably had their origins back in the Bronze Age or earlier. The tinners' year was broken up by regular festivals. They celebrated the start of the year and the end; the time each year when they brought their black tin to the bowl furnace or blowing-house for smelting; and they memorialised the discovery of tin and smelting by their dimly remembered forebears, discoveries that later became associated with St Piran. These occasions were noisy and outrageous celebrations. Some of these festivals were still observed in the second half of the 19th century, and in some parts there remains a sort of folk memory even now.

The early tinners' isolation from other folk meant that they were able to keep up their customs and traditions over many centuries. But as larger groups worked in increasingly organised mines, and with constant contact with 'foreigners', it was inevitable that these ancient practices should be modified and eroded until their unique features were compromised. Now, only the famed humour and independence of the Cornish miner remain to remind us of their former way of life.

As the domination of Cornwall by Wessex changed, and it became part of the realm under King Athelstan (AD 924–39), the tin industry continued to be significant in the local economy. West Saxon government administration was extended across Cornwall as their system of land tenure and justice was imposed on the Cornish. The 11th century saw the situation change as increasing raids by Danes and, after the Norman Conquest, by disaffected Saxons disrupted the tin industry, leaving us with a picture of occasional and fragmented activity.

CHAPTER TWO

PRIVILEGE AND REBELLION

Medieval period (1066–1485)

OPPOSITE
An aerial view of Ding Dong Mine. The lodes of this district have been worked since prehistoric times and physical evidence of outcrop workings from medieval times can be seen here, as well as workings from the 18th and 19th centuries, like the engine house at Greenbarrow Shaft, seen in the foreground.

AFTER THE CONQUEST, the comparatively advanced and well-organised administration of the Saxon rulers was continued with little alteration by the Normans. Military and ecclesiastical arrangements apart, the actual organisation of the government, as well as its taxation and coinage regime, continued much as before, at least until the reign of Henry I (1100–35). The great economic survey known as the *Domesday Book* emphasised continuity rather than innovation in that, when giving details of land ownership and values for 1086, it referred back in each assessment to the position during the reign of Edward the Confessor, the last Saxon king the Normans recognised. Domesday covered the whole kingdom of England, from Cornwall to the far north, and despite the detail it goes into on all aspects of the land, its value and who and how many worked it, it makes no mention of tin or tin works. Iron and lead mines elsewhere in the country are mentioned, but not tin. As tin was being worked at that time in both Devon and Cornwall, various explanations are given for this omission. G.R. Lewis thought it probable that the industry was either temporarily granted to a great magnate or that it was was held by the crown and hence was not assessed.[1]

Although there is little indication of how mining and ore dressing might have progressed at this time, we do have information on the number of water-powered grist mills operating in the southwest. The *Domesday Book* indicates that there were only six corn mills in Cornwall, and two of those were on the border with Devon. The fact that this technology was far commoner in the east of the country than in Devon and Cornwall might suggest that ore crushing and grinding machinery were unavailable at this time and that water-powered bellows for blowing-houses were also introduced at some later time.[2]

Reorganisation of the tin industry under the early Plantagenets

At the end of the civil war of 1135–54 the Angevins (Plantagenets) came to the throne and the fiscal arrangements of the government were tightened. After Henry II (1154–89) took control his officials began the records known as the Pipe Rolls. These contain the first records of tin production, and although the earliest refer only

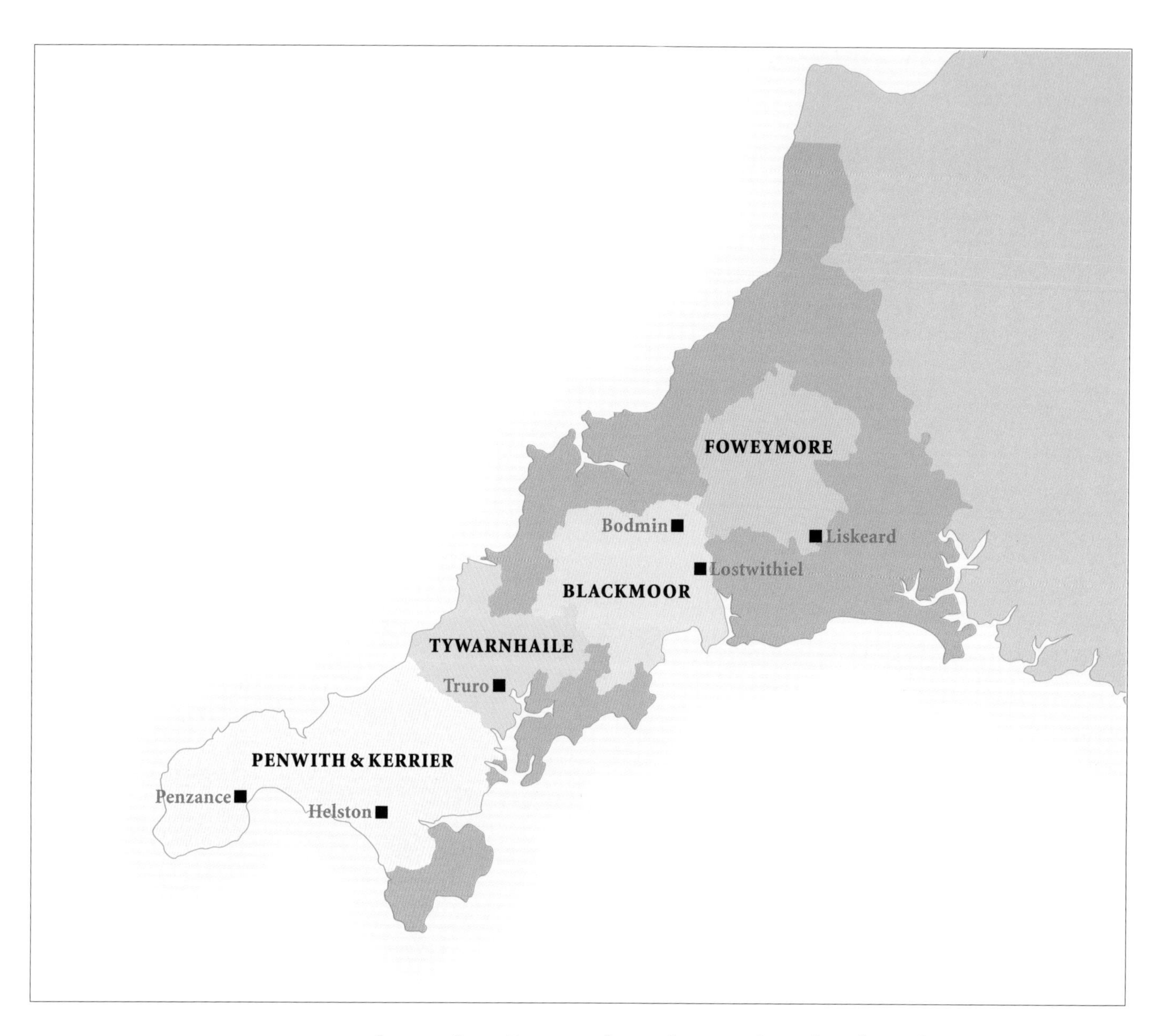

The map shows the four stannary districts of Cornwall and their coinage towns. Lostwithiel was the administrative centre, with the Duchy Palace and the tinners' prison. Penzance became a coinage town in the 17th century.

to production from Devon, other indicators show that the industry was quite productive in both counties. It was nearly 40 years before Cornish tin was mentioned in the rolls, in 1195, and the tonnage then was shown to be less than half that from Devon. With information from the Pipe Rolls and other sources, Angevin officials were able to examine the potential of the tin industry. Henry II's two sons both contributed to the subsequent process of reorganising the stannaries. In 1201 King John (1199–1216) granted the first stannary charter to be issued by the English crown. This was the second move in an overall plan to bring the Cornwall and Devon tin industries under closer control and thereby increase royal revenue. Three years before John's charter, the justiciar of his older brother, Hubert, Archbishop of Canterbury, who exercised authority in Richard I's absence, began the process of enhancing and stabilising the crown's income from various duties and taxes.[3]

Hubert appointed William de Wrotham to be the first warden of what became known as the 'stannaries'. It appears that before his appointment the sole connection between the crown and the tin industry was taxation. The organisational arrangements of the industry before 1198 can only be guessed at. Some historians make connections between the Roman period and the early Medieval period, but these are at best tenuous and optimistic. The Saxons came as raiders, stayed as conquerors, and created institutions which were in large part based on their own customs and practices. As Christianity influenced them more they sometimes adopted arrangements common to other, more advanced, European countries. Roman laws and practices may also have influenced them through these connections. What is also evident is that the older inhabitants of the tin-producing districts of Devon and Cornwall – who had undoubtedly continued as tinners throughout the turbulent times that followed the Roman withdrawal from Britain – also had customary practices that survived until the time of John.[4]

When de Wrotham took over the tin industry, which had long been centred on Dartmoor and east Cornwall, its centre was already moving westward. In the 12th century Bodmin appears to have been the most important coinage town (one where the smelted tin ingots were taken to be assayed), and its tin was exported to the continent through Lostwithiel, the nearest navigable water. Bodmin and the stannary of Blackmoor remained the most important part of the industry throughout the 13th and first half of the 14th centuries. Merchants from southwest France, particularly Bayonne, are recorded as having purchased tin at Bodmin in that period, and two merchants, Auger de St Paul and William Père from Bayonne, bought 128 tons of tin there in 1198. An Arab traveller of the 12th century described the route that Cornish tin took as it was exported via France to the Mediterranean. It was taken to Bordeaux and then by boat up the River Garonne to Toulouse and by pack horse to Narbonne, from whence it was carried by French ships to Alexandria and the rest of the eastern Mediterranean. Marseilles was also an important entrepot for the distribution of Cornish tin, just as it had been thirteen hundred years before, in the time of Pytheas. The failure of the Pipe Rolls to record much of this twelfth-century activity indicates that most of it went on unregulated and untaxed and that the tin was probably smuggled out of Cornwall.[5]

William de Wrotham's appointment was intended to rectify a situation which the crown thought unsatisfactory. The letter outlining his appointment, powers and responsibilities helps us to appreciate the very rudimentary arrangements for controlling the industry that

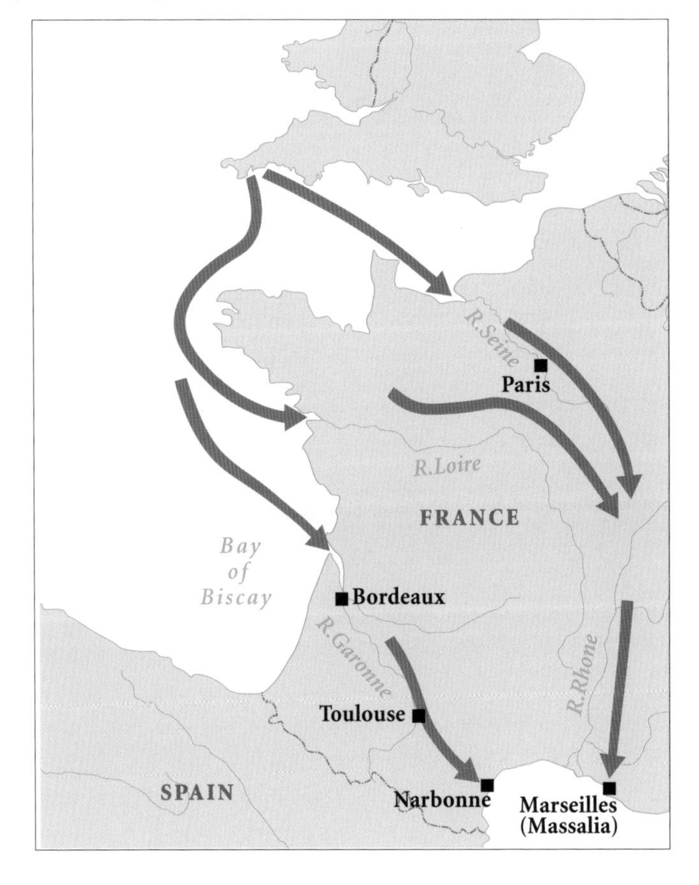

This trade route map shows that the principal ancient markets for Cornish tin were the civilisations around the Mediterranean Sea, so the trade routes went through Gaul or France to Massilia (Marseilles) and other ports in the area. In medieval times trade increased to northern, central and eastern Europe.

had existed hitherto. De Wrotham was granted wide-ranging executive management powers, as well as the legislative authority to make new rules and regulations as he thought necessary. These pertained to how the tinners were to work, how they were to be controlled, how and under what circumstances tin was to be smelted and weighed, and details of the new duty and taxation regime. Most of the innovations related to taxation, and a new tax of one mark (13s 4d) per thousand weight (1200 lb) was introduced. De Wrotham summoned juries that represented tinners from both Devon and Cornwall and whose main function appears to have been to make tax gathering easier. These appear to have been the earliest 'stannary courts'. An important aspect of the new regime was concerned with the regulation and standardisation of weights and measures, along with the means and personnel to check their accuracy. The new warden was able to appoint a whole range of petty officials, including collectors, controllers and treasurers. The result of this reorganisation was a new, stricter code of practice that ensured a better, tighter control of tinners and their industry; and, of course, it also meant a more stable and reliable flow of cash into the exchequer.[6]

The stannary charter of King John (1201)

Tin production figures from 1156 to 1200 show a steady rise, and although they are only partial – in that for the most part they do not include production from Cornwall – they do indicate that the government officials in charge of taxation were gradually getting a measure of the industry. The sudden drop of some 10 per cent between 1199 and 1200 probably had more to do with lack of control and evasion than the real situation at the tin works. It may be that this situation was what led to the issuing of the first stannary charter by John in 1201. Despite King John's well-known failings he has been described as a good administrator, and he was astute enough to see that de Wrotham's system needed tidying up so that the tinners were not merely controlled but also encouraged. John continued the policies of his father and brothers, but he sought to extend and manipulate them by clearly defining the liberties of the tinners and granting new controls to his 'custos' or warden. His great 'Charter of Liberties to the Tinners of Cornwall and Devon' was a subtle mixture old and new, of continuity and innovation, of carrot and stick, of advantage to the lowly tinner and a well-controlled, reliable source of revenue for himself. The very first clause offered an important inducement to would-be tinners: 'We have granted, that all our tinners in Cornwall and Devon shall be free, and quit of plea of natives, whilst they work to the advantage of our farm.' They were to be outside many of the ordinary restrictions and impositions of the feudal system. As part of the king's demesne the tinners came under a royal official and enjoyed fairly comprehensive protection from the action of sheriffs and other county officials. This assertion of the crown's rights over the stannaries emphasised that tin, like other metallic minerals, was owned by the crown.[7]

The second clause granted the tinners four liberties: 'Without disturbance of any man' they could 'dig tin … turves to smelt the tin … buy wood for the smelting of tin … [and] divert waters for their works in the stannaries.' Moreover, these liberties allowed the tinners to do such things even on the land of bishops, abbots and earls. As noted in Chapter 1, John's charter stated that these were ancient rights which he was confirming, and he reminded both tinners and feudal magnates that he was merely regularising established practices and customs for the mutual benefit of both parties, rather than introducing onerous innovations. Thus, he gave a legal framework to long-established practice.

The next clause placed the tinners under the jurisdiction of the warden and his officials. Sheriffs and other royal officers would have no power to summon tinners to their courts. The charter gave to the warden all power to judge matters in the stannaries and to protect the privileges of the tinners. There would be a stannary prison, and even if a tinner became an outlaw he would still be subject to stannary law as exercised by the warden. Finally, the charter gave specific protection and privilege to petty officials, such as weighers and treasurers, who were to be exempt from taxation in local towns and markets.[8]

The reference to the right to work in the lands of bishops and abbots is appropriate, for religious houses like the Benedictine monastery on St Michael's Mount and Tywardreath Priory held land that contained tin works. For example, in 1170 an agreement was made between the prior of Tywardreath and Philip de Treverbyn concerning workings at Levrean, near St Austell, in the stannary of Blackmoor. The agreement hints at customary rights, but it also sets out a locally agreed separation of rights whereby the farm of tin (the right to work it) was divided between the priory and the tenant. John's charter was to regularise such arrangements. Eighty years later, in 1250, Philip's descendant Odo handed his half of these tin works back to the priory. The cartulary (collection of records) at the monastery on St Michael's Mount also has reference to a tin working in St Hilary parish, more than 30 miles from Levrean, and which was also called Treverbyn.[9]

John's charter did not immediately get the desired results, and it was nearly a decade before production showed the hoped-for increase, rising from some 432 tons in 1201 to a peak of 642 tons in 1214. However, these Pipe Roll figures tell us only part of the story. A more significant development was the reaction to the charter by the barons and feudal landlords. They resented the loss of power and control over the lowly tinners, who, taking full advantage of their recently confirmed rights, no longer felt themselves subject to their erstwhile feudal overlords. With John under pressure from his barons, the liberties granted to the tinners in 1201 were effectively withdrawn just before Magna Carta was forced on him in 1215. John assured the landowners of Cornwall and Devon that they 'should not lose by reason of the stannaries aught of the services or customs which they are accustomed to have from their men and serfs', although he also made clear that their men were still allowed to 'go for tin'. Fortunately, during the reign of Henry III (1216–72), the original charter was confirmed in all its details and was to continue in force until Edward I issued his great charter in 1305.[10]

Patterns of trade in the 13th century

During the early years of the 13th century tin production and export settled down into a pattern that suited both the crown and the tinners. Despite the concerns of the feudal lords, production continued and, although the only year before 1301 for which the Pipe Rolls give figures for Cornwall is 1195 (137 tons), there is ample evidence that tin continued to be exported from all the principal stannary towns. Much of it was bought by French merchants, mostly from Bayonne, Bordeaux, La Rochelle and Oléron, and by London pewterers. Their agents visited the stannary towns and arranged transport. In 1265 a request was made on behalf of the burgesses of Exeter, Bodmin, Truro and Helston for 'safe conducts' to enter Bordeaux and La Rochelle to sell their tin. How usual it was for Cornish merchants to sell their tin abroad as opposed to dealing with visiting merchants we do not know. However, this does tell us that both Truro and Helston were becoming significant producers and that the burgesses of those stannary towns were involved in the tin business. Another rich market was northern Europe, and merchants of Brabant purchased tin from Cornwall and Devon in the 12th and 13th centuries, as did Hanseatic agents, who took tin to Bruges, Houke and Oostkerke for export on to all parts of north, central and eastern Europe. The great cities of Lübeck and Hamburg sent an agent to Falmouth to secure a steady supply of tin in the 1260s. The discovery of tin in Bohemia in the 12th century failed to inhibit European demand for Cornish tin, and it was the 14th century before the southwest's monopoly was affected by it.[11]

Edward I's Charter of Liberties (1305)

By the end of the 13th century tin was an important source of financial support for the crown, and as Edward I (1272–1307) pursued his wars against the French, Welsh and Scots – and built up large debts – it was revenue from Cornish and Devonian tin that helped to bail him out with his creditors. In 1297 Edmund, Earl of Cornwall, loaned the crown 7000 marks to pay off its debt to the merchants of Bayonne. Though this huge sum amounted to two years' revenues from the stannaries it did not solve Edward's debt problem, and he also had to farm out silver mines in Devon to Italian financiers to ease the pressure on his exchequer. It was at this time that the wealthier classes of the stannary districts of Cornwall moved to take greater control of an industry which was proving lucrative. Seven gentlemen from the Blackmoor stannary, mainly centred around St Austell, persuaded the government to issue a new, more comprehensive and wide-ranging charter. Thomas Beare, in his book *Bailiff of Blackmoor*, written in 1586, gives us the background to the issuing of Edward I's *Charter of Liberties to the Tinners of Cornwall* in 1305. After describing the banishment of the Jews from England in 1291, Beare refers to the 'tynners having great profyts by their tyn', which he thought they 'wrought … by custom'. To secure the continuance of this profitable business, these seven gentlemen needed the crown to grant a charter that would guarantee their rights and give them more control of their industry through the establishment of a proper court system. De Wrotham had allowed for a rudimentary court and, although John's charter does not mention one, warden's courts were operating regularly by the 1240s. Edward's charter set out who were to form the juries in these 'inquisitions', or courts. If a case involved a tinner but had nothing to do with the stannaries, half the jury would be tinners and half 'foreigners'. If the case involved only tin and tinners, then the court 'shall be made as they have heretofore been accustomed', with only tinners as jurymen.[12]

This tin-zinc figurine of a seated man was found 3 m (10 ft) below the surface on Bodwen Moor, Lanlivery, in 1853. Various explanations have been offered as to its origin, but it is thought likely to date from the late 13th century, and to have been connected to Jewish involvement in the tin trade.

There is a common belief that Jews were involved in the business of the Cornish tin industry, either as financiers or as managers of actual tin works, in lieu of the crown's debt to them. Beare stated that the Jews played no part in the stannaries, living as they did in the big cities rather than the country. However, support for their involvement comes from a tin–zinc figurine found 10 ft (3 m) below the surface at Bodwen, Lanlivery, in 1853. This seated figure is thought to have Hebrew lettering on it, and various interpretations have been offered. If this is true, it might lend credence to the idea that Jews were involved in the industry before their expulsion from England in 1291.[13]

Edward I's charter of 1305 cleared up some problems with John's and added details that his predecessor's had omitted. The warden would not have total control of the tinners, for the county coroner would also have power to intervene. A fugitive tinner could subsequently be taken before the 'justices itinerant in the county'.

The charter also named the stannary or coinage towns which were to be used for the control of tin and the collection of duties. These towns – Helston, Truro, Bodmin, Liskeard and Lostwithiel – had long been the established centres of their respective stanniferous districts, but the 1305 charter limited coinage to these places by law. Edmund, Earl of Cornwall, who controlled and received the income from the stannaries from 1272, built what has become known as the Duchy Palace at Lostwithiel. Edmund's hall there is referred to in a deed dated 1292, and it seems that already administration of the tin industry was centred on the earl's palace at Lostwithiel.[14]

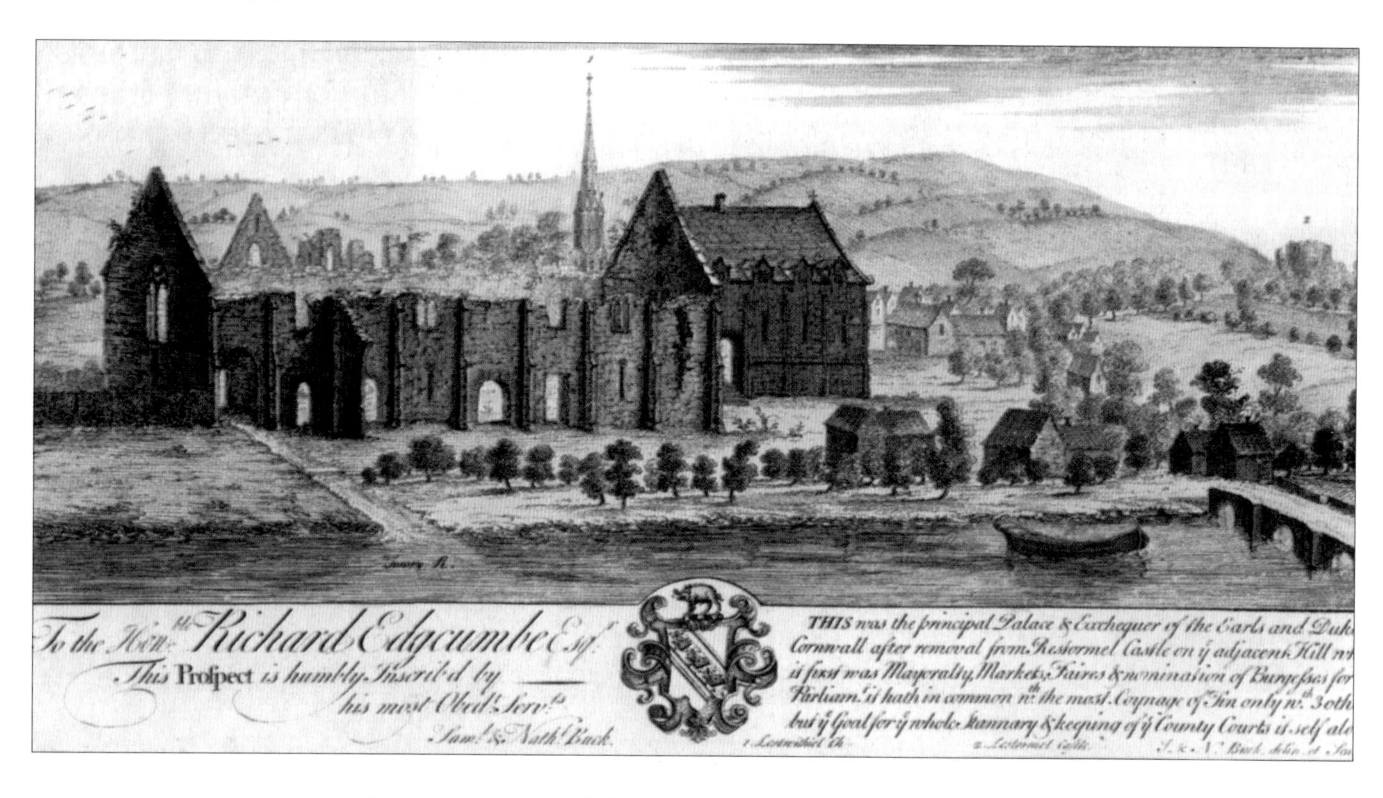

The Duchy Palace at Lostwithiel. The building was there in the 1290s, as the hall of Edmund, Earl of Cornwall, who was responsible for the stannaries. It had become the administrative centre for the whole of the Cornish stannary by the time of the creation of the Duchy of Cornwall in 1337. The building was badly damaged during the English Civil War.

Management of the tin industry in the 14th century

After 1305 the ancient customary practices of the tinners, already partially regularised by de Wrotham and King John, became a well-organised and thoroughly regulated system. The stannary districts were Penwith–Kerrier in the far west, with its coinage town of Helston; the small stannary of Tywarnhaile between Truro and St Agnes, with its coinage town of Truro; the large central stannary of Blackmoor, with coinage at Bodmin; and the most easterly stannary of Foweymore, consisting mostly of moorland north of Liskeard, the coinage town. The hub around which the whole system operated was Lostwithiel, with its palace. There the principal officers were based, and the stannary prison was also located there. The management was simple and to some extent typical of medieval arrangements. The lord warden, or chief warden, was always a great magnate and usually a court favourite; hence, the job was a sinecure. The man who actually ran the stannaries was the vice warden, and he would normally be a Cornish gentlemen of note with an interest in the tin industry. He acted for the lord warden in all matters and presided over the Great Court, where matters of importance were settled. This sat only occasionally, and the jury was made up of 24 tinners, six from each coinage town. They were appointed by the mayors of these towns. Below the vice warden were the four stewards who represented the four stannary districts. They had jurisdiction over their own stannaries and power to call and preside over stewards' courts, which had developed since the 13th century to deal with day-to-day problems in the stannaries. These courts met usually about every three weeks and dealt with disputes between tinners over bounds, the registration of new bounds, problems arising from the diversion of water from one works to another, and other infringements of customary practice. The stewards' courts were itinerant, moving around the tin districts and making use of the houses of the gentry and local inns to conduct their proceedings. Examples of the way that the circuit operated can be seen by examination of the records for the Penwith–Kerrier stannary. Between 1356 and 1515 the court sat at Mousehole, Alverton, Marazion, Goldsithney, Lelant, Treslothan, Redruth and Helston. Issues not settled by a steward's court could be referred to the Great Court for a decision. This usually sat at Lostwithiel but probably moved to coinage towns like Bodmin or Truro on occasion.[15]

The steward's right-hand man was the bailiff. He was the man who actually supervised the smooth running of the tin workings. His was the task of visiting the hundreds of individual bounds spread across dozens of moors and hillsides. The bailiff had to be familiar with every set of bounds, every controversial 'side bound', every water supply, whether stream or leat, every set of stamps, all the blowing-houses, and the entire system of security involved in the carriage of white tin to the coinage town and its safety once there. The silting up of fresh-water rivers was his concern, as were adequate forage and grazing for the horses of charcoal burners, and the correct measurements for charcoal sacks and the weighing of black tin. The bailiff had to ensure that the blowing-house's and individual blower's marks were correct and that the white tin was assayed accurately and given the appropriate quality mark. When a tinner needed arresting, the gaol-keeper disciplining, a debt collecting, a writ serving, a petition presenting, a violent dispute settling or 'shovell money' collecting, it was the bailiff who had to do it or see that it it was done. Failure to perform any of the above laid him open to arrest or fine. He also had to ensure that all bounds were properly registered and that the registration was legitimate.[16]

Petty officials working under and answerable to the bailiff were the tollers, porters, peisers, comptrollers and receivers. The toller was a sort of assistant bailiff, the porters moved the heavy blocks of tin metal and the peisers checked the weight

OPPOSITE
Leaden seal found near Bath, Avon, in 1842. The seal carries a Latin inscription which translates as, 'Seal of the Community of Tinners of Cornwall'. The seal was probably attached to the Edward I charter of 1305.

of each block, using the 'king's beam' and officially approved weights based on the standard set kept in Winchester to check the weights. The comptrollers were in charge of discipline and security, and the gaol keeper came under them. The receivers looked after the stannary accounts, ensuring that the books were in order and that the correct duty was paid. Most of these activities took place at Lostwithiel. Coinage itself, carried out in each coinage town, consisted of cutting off a corner (*coin*) from each tin block for assaying. Most of these positions, duties, organisational arrangements and definitions changed and developed over several centuries.[17]

New technology and improved tin streaming and metal mining

In the centuries following the Norman Conquest there were many changes in the way tin was won and processed. Knowledge was gained about more efficient ways of locating tin deposits, better methods for the extraction and separation of ore were developed, and the materials used to make tools were improved. Water-powered machinery also came into use, and with better organisation and the involvement of wealthier tinners the industry became more efficient. By the early 14th century water-wheels were powering bellows in the smelting process, creating blowing-houses. The tinners were forced to search for less accessible and lower-grade tin in the stream workings, and this led to greater sophistication in dressing the ore and increased care to ensure that little was lost. Tinners learned new techniques from each other and modified and adjusted their use of ancient methods.[18]

As the available alluvial tin was worked out, the tinners turned to the primary source, which was the lodes. Mining a lode was a different job to streaming alluvium, and several technical problems had to be overcome to work safely and successfully. At first, as with the exploitation of surface outcrops in ancient times, the tin lodes were worked in long trench-like excavations that were open to the skies. These 'coffin' works were sunk to a considerable depth, until either the dip of the lode (the angle of the upper surface of the lode to the horizontal) or the sheer impracticability of hoisting ever-increasing tonnages of overburden from greater depths caused the miners to turn to the method known as 'shamelling', which involved working down on the back of the lode in large steps, with the ore being thrown up from step to step to surface. This inefficient method soon gave way to conventional shaft and level mining. There are no records of underground tin mining from the period, but we do have some clear descriptions of lead–silver mining in the royal mines on the Cornwall border, at Bere Alston, and on the north Devon coast, at Combe Martin. The records start in 1292, and for the next few years there are references to all aspects of mining, mineral processing and smelting. Leather buckets were used to hoist the water that came into the mines, and they also appear to have been used for drawing ore to surface and for lowering materials underground. Tallow candles were used for lighting and iron for the tools, which included poll picks, hammers, chisels and shovels. The points of the picks were steeled to make them harder and last longer. Heavy timber was also needed to support the workings, as well as lighter wood for ladders, shaft casings and a host of other functions.[19]

These records from the Bere Alston silver–lead mines tell us more than just about the mundane day-to-day running of the mines and the expenses. They also reveal that there was progress in mining techniques, ore processing and smelting. With the bringing in of miners from Derbyshire, the Mendips, Wales and Cornwall, the problem of coping with water ingress was solved. Some of the Cornish tinners came from Alternun, on Foweymore, and undoubtedly brought with them skills that had been developed there over many centuries. By the late 1290s adits (tunnnels) were being driven into the workings from lower down the hillsides to give the water a natural, gravity-driven drainage outlet, obviating the need to bail with leather buckets. By 1297 a system of drainage levels had been introduced into all four lead–silver mines working at Bere Alston. The need for efficient ventilation was also recognised and dealt with. On the surface improvements were made to the ore-breaking process through the introduction of 'breaking black ore [galena] with machines' (4 June, 1306) and 'breaking black ore with engines [using] 19 horses'(13 August). Peter Mayer thought the machines (engines) were crasing mills turned by draught horses, but equally they may have been a form of stamping mill. The smelting process was also becoming more sophisticated. The primitive but fairly efficient bole, which used a natural draught to increase the temperature, was supplemented by various hearths operated by skilled refiners. Workers operated bellows with their feet to blow air into the furnace. The record for May 1306 refers to a 'certain new device made which is called a "terengle" to smelt ore'. In August there is a reference to a 'furnace man … making a new device.' There is also mention of water-powered bellows being used at the smelting furnaces at Maristow. What all this tells us is that mining and ore crushing, processing and smelting were being constantly improved in the southwest, and that the tinners of Cornwall were both contributing to and learning from it.[20]

This iron-shod tinners' shovel is a typical example of the tool described by Richard Carew in 1602, as a 'broad shovel, the utter part of iron, the middle of timber, into which the staff is slopewise fastened'. This example was found in a tin stream on Luxulyan Moor and can be seen at Truro Museum.

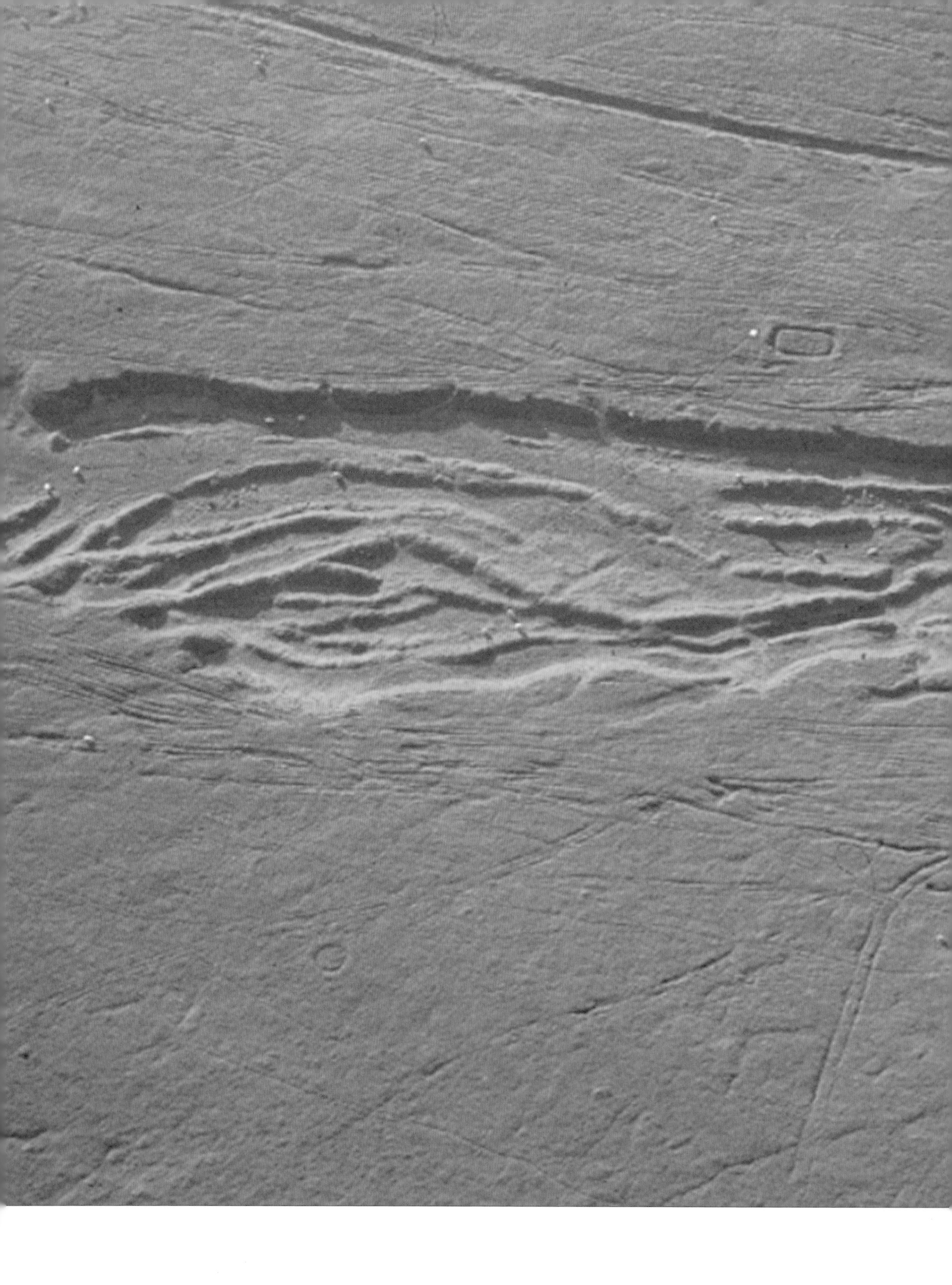

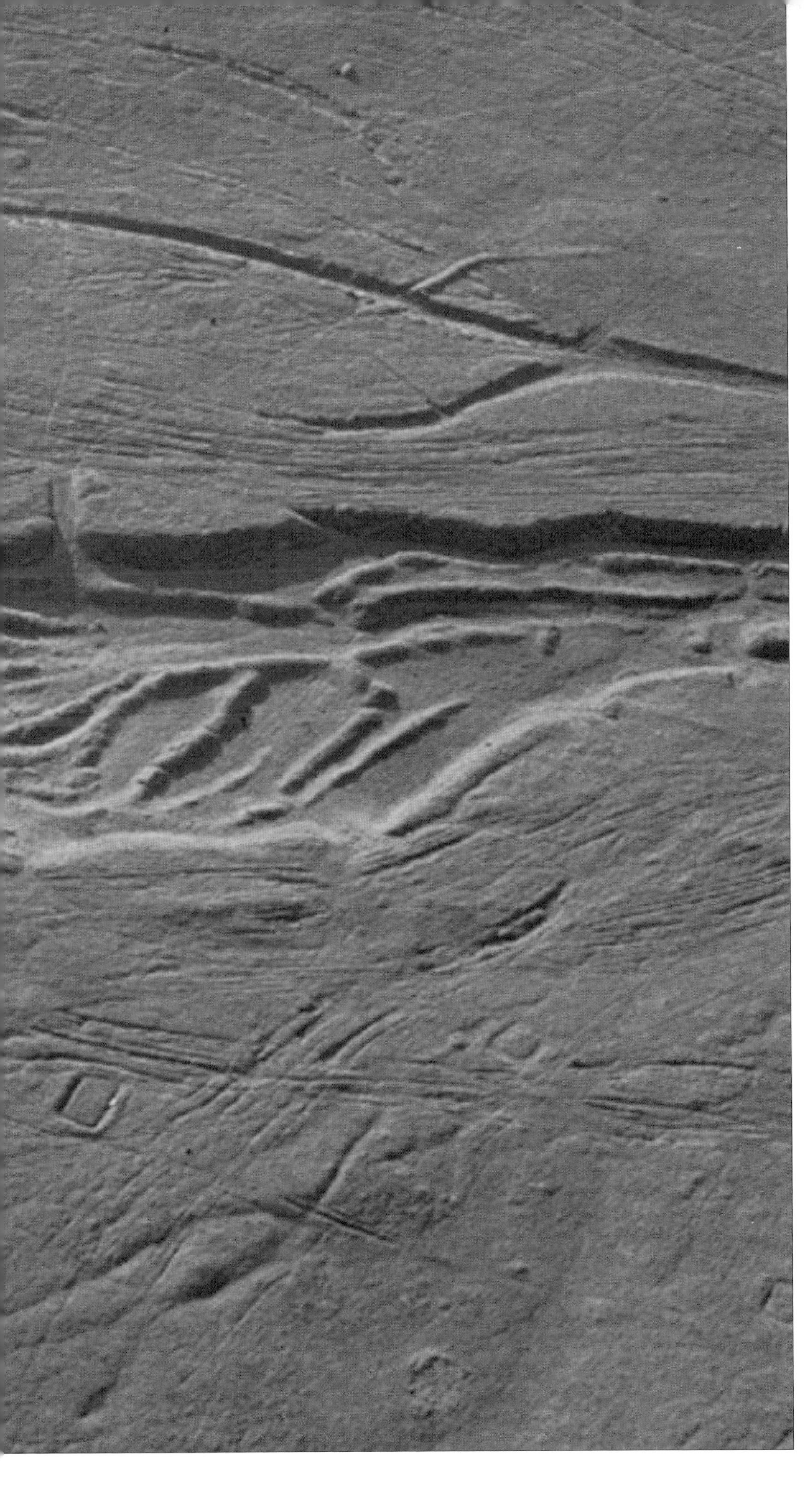

An expansive aerial view of West Moorgate eluvial stream working on Bodmin Moor (Foweymore). Eluvial stream works exploited cassiterite detached from the lode outcrop by weathering and deposited downslope. This eluvium or 'shoad' usually lies on hillsides in shallow, fairly dry valleys above rivers. Here, stream-work tinners have removed the overburden, causing curved dumps. The angle of dumps relative to the slope of the hillside depends on the hill's steepness, as the velocity of water flow is crucial in the separating process.

The status of tinners and the structure of tin workings

The status of tinners during these centuries is of interest because they had long enjoyed a uniquely free and independent existence. The position of the miners under the Barmoot system in Derbyshire, or of the 'free miners' of the Mendips and the Forest of Dean, was similar in many respects to that of their counterparts in Devon and Cornwall. The tinners were largely exempt from many of the controls and restrictions under which most of the labouring poor suffered. The charters of kings John and Edward confirmed ancient privileges and liberties, which left the tinners able to get on with their lives without undue interference from local officials and lawmen. They were exempt from most taxes, duties and impositions. And if their poverty was a byword, they were relatively free. In the 1580s Beare commented that despite the tinners' legendary poverty, people envied them their independence and self-sufficiency, and in particular they admired their charity towards each other. It took the horrors of the Black Death of 1348–49 and the ensuing sudden shortage of labour to bring such independence to most of the English peasantry. The financial set-up of the tin works meant that those who had a share in a set of bounds enjoyed a share of the profits. They also, of course, had to meet their share of the costs.[21]

The unique status of tinners and their exemption from various impositions continued to cause problems and disputes, and these came to a head in the second half of the 14th century when the definition of precisely who was a tinner was taken before Parliament. In 1376 the 'Commons of Cornwall and Devon' sent a petition to the 'Good' Parliament seeking to restrict involvement of the wealthy in tin works and to ensure that only working tinners enjoyed the privileges accorded to them. Parliament responded by defining tinners as those who worked manually in tin works, a definition that was confirmed by Richard II (1377–99) some time later. Although this remained the official position until the early 16th century, it was largely ignored in that wealthy entrepreneurs and merchants continued to operate as tinners under the protection of the stannary laws.[22]

During the late medieval period there appear to have been different forms of organisation in place in the tin industry. This structural diversity was to continue, to some extent, right into the 20th century. Individual 'free miners', working their own bounds, sometimes full-time and sometimes as by-employment, have probably existed since before records began, as have the commoner partnership arrangements. Finance was always a problem for the lone tinner as well as for associations of tinners. Lack of finance inhibited the search for new tin deposits and left tinners vulnerable when the ground they were working became exhausted or when violent storms destroyed their tin works. From the very beginning, however, the wealthier classes were involved in the organisation of the tin industry. At first this was probably as buyers of black tin and controllers of the smelting arrangements, but undoubtedly they later used their money to become fellow adventurers or shareholders alongside the working tinners.[23]

Many of the smallest tin works, such as those in a stream or moor at the bottom of a farmers' field, were worked like any other crop as part of the farmer's exploitation of his land. Until the 17th century it was common for wills and inventories to record the tools 'on the moor', such as shovels, picks or wheelbarrows. These streams were worked in season – when the planting or harvest was done and there was sufficient water flowing. The small associations of tinners worked on a more regular basis, and among their number would be skilled full-time tinners and casual wage-earning labourers. Wealthier adventurers or entrepreneurs, with shares in several tin works, would send 'spaliers' to stand in for them. A spalier was a day worker who worked in lieu of the 'spale' or fine owed by a shareholder for non-attendance at the working. Each shareholder was responsible for an equal share of the cost of the tin work and the time spent working it.[24]

To the entrepreneurs who were involved in several workings could be added other sleeping partners, such as money-lenders and merchants, who had acquired shares in exchange for taking on debt. These helped to widen the gap between investors and working tinners, and between capital and labour. Tin works and shares in them could also be leased in medieval times, and the charters allowed this. A share in a tin work was a chattel and could be sold or willed to family, friends or the local chantry priest or parish church. Camborne church was bequeathed a share in Carnkye Bal (Illogan), and Gwennap church was left a share in Poldice mine (St Day). Richard Carew, in the 1580s, described the formation of associations of the wealthy to spread the cost of a new mine, but the practice was already centuries old. When Prince Edward granted Henry Nanfan exclusive rights to the tin at Lamorna in 1359, Henry had 15 partners. There were 48 partners in a Blackmoor tin working, all of whom were listed on a deposition as owing tin to Michael Treneweth. An exception was Abraham the Tynner, who appears to have been the sole owner of some seven tin works, including at least one lode mine, in the 1350s.[25]

Legal disputes in the 14th century

As tinners became more active all over the stanniferous districts of Cornwall, references to tin works and complaints about their activities and the resultant damage to property increased. In 1301 a complaint was brought against tinners working close to the bridge of Reswythen, between the parishes of Redruth and Illogan, below St Uny Church. During the last years of the 13th century Ralph Wenna and John de Treveyngy, both tinners, had apparently undermined the foundations of the bridge with their tin works. They denied the charge, but they were found guilty and their goods were seized to pay for the bridge's repair. In February 1352 there were complaints to the authorities about damage to the water supply to Lostwithiel corn mills by tinners working upstream at Glyn and Redwith. The same complaint came up again in 1357. Tinners were constantly accused of causing the harbours of Fowey, Plymouth and Falmouth to silt up as a result of their activities. In July 1353 a serious complaint was made against Sir John Beaupre, who was accused of employing armed men to attack a tin work at Marazion, kill one of the owners, wound another and drive off the other tinners. The gang burned down the petitioner's house and stole his tin. In 1357 there was a long-running dispute between local officials and Abraham the Tynner, who owned half-a-dozen tin works in Foweymore stannary. Abraham and his fellows were carted off to Lostwithiel gaol and held prisoner. Most of his workings were stream works, but at least one was a lode mine, situated in the Glyn Valley. Abraham employed 300 tinners in his works, and he pointed out to the court that he had handed large profits to the Black Prince through tin duty. Officials appeared to be taking advantage of the prince's absence in Gascony to grab what profitable tin workings they could, but they were not as efficient as the tinners and the prince was worried by the interruption to his money supply. Later in 1357 the parson of St Ladock accused tin miners of undermining his graveyard and destroying trees on his property. He specifically said that the prince's absence in Gascony encouraged the tinners to act outside the stannary laws, which forbade working under church property. In 1359 the prince granted exclusive rights for a period of 12 years to a group of wealthy men to work the moor at Lamorna. This began a series of events which led to several armed invasions of the tin works and one of the longest-running court battles in medieval tin-streaming history. In 1361 sixty tinners were accused of diverting a stream across good agricultural land and destroying wheat, barley, oats, hay and peas growing there. All the topsoil was washed away, rendering formerly productive land sterile.[26]

These examples show that significant tin production was taking place in all the stannaries of Cornwall and not just in the central Blackmoor Stannary. In Gwennap, for example, a tin working called Stence Segh continued to be productive for over 200 years, between 1327 and 1530, when it was known as Stagnar Segh. Stannary records for the year 1305/6 show that just three wealthy Gwennap tinners between them produced a total of more than 100 tin ingots weighing over five tons that year – a considerable achievement![27]

Other developments in the 14th century

The 14th century saw many changes in the tin industry, with production rising to new heights, the creation by Edward III of the Duchy of Cornwall, and the devastation of the Black Death. In 1301 Cornish tin metal production was recorded as 560 thousand weight (300 tons), but by 1338, the year after the Duchy came into existence, it had risen to 1228 thousand weight (658 tons). Revenue to the crown from tin duty rose from about £2000 a year to £3000 in the late 1330s. Pre-emption – the right of the crown, and later the Duchy, to buy the whole output of tin – had been exercised to a varying degree for generations, but with the wars of Edward I, Edward II and Edward III tin became an easily used commodity to pay off creditors. Edmund, Earl of Cornwall, had used his right to great financial benefit, storing one year's tin, selling it for a high price and then buying the next year's tin from the hard-pressed merchants cheaply. Edward I borrowed from the merchants of Bayonne. Edward II used the Genoese banker, Pessano, securing loans by handing over revenue from the stannaries and granting him a lease of tin pre-emption. By the 1330s the great merchant bankers of Florence, the Peruzzi and Bardi, were involved in financing the English crown, and soon they were exercising control over the export of tin. With the creation of the Duchy in 1337 tin revenue went to the seven-year-old duke, Edward of Woodstock (Prince Edward, known as the Black Prince), and by the time he was sixteen – at the time of the Battle of Crécy (1356) – he was drawing a sizable income from the stannaries. The middle years of the 14th century were times of great difficulty. As we have seen, Cornwall was riven by lawlessness, and during Edward's tenure of the dukedom things did not improve. He constantly complained of loss of revenue from the stannaries due to illegal invasions of tin works, violent disputes between tinners and, worst of all, the Black Death. In 1359 the duke moaned: 'The stannary of Cornwall has greatly deteriorated and is now of much less value … owing to divers disputes among the tinners … and partly owing to the lack of workers since the pestilence.' The disease arrived in England in 1348 and devastated the whole country, killing between a fifth and a third of the people. The first figures for tin production after the outbreak, in 1357, show a drop of over 50 per cent from 1338, and whole areas of tin works were abandoned as there was nobody to work them. By the end of the century production was back to the level of the late 1330s, and, although population numbers did not recover for a considerable time, the tinners' output did, possibly because newcomers were drawn into the industry. When Henry IV (1399–1413) took the crown from Richard II he reissued Edward I's charter, wording it in such a way as to emphasise that everything was to continue as before.[28]

The 15th century – a time of change

The 15th century witnessed the continuing move westwards of the centre of the tin industry and also the more significant move towards lode mine working. By the middle of the century underground mining was well established, and the proportion of tin these mines produced was also rising. Despite the widespread lawlessness, which affected the whole country for a large part of the century, causing problems in every county and most industries, tin production was maintained. Tonnage dropped in the first half of the century to half of its total in 1400, but despite significant fluctuations it had largely recovered by 1495. Violence in the tin works became endemic as gangs, often led by local gentry (who frequently were also justices of the peace!) badly affected production. But it was also a time of technological advance, as water power was applied to machinery for several purposes. Not only did the use of blowing-houses become widespread and more sophisticated, but water-powered crazing mills also appeared in most districts. The earliest reference to water-powered stamps was in 1493, at Treneer Wolas, in Wendron. Within a decade or so these powerful and efficient machines were being used wherever lode ore needed crushing. Within 25 years their use had been exported to Germany and beyond. Water-powered pumps were coming into use on the Continent, and although we have no evidence that they were used in the tin mines of the 15th century, they appear to have been employed on the Tamar silver mines at that time. According to Lewis the 'ordenaunce' that was used to dewater the Beer Ferris mines in 1480 was a small water-wheel and suction pump. It is hard to believe that this technology was unknown just a few miles away in the tin mines of Cornwall.[29]

OPPOSITE
The drawing details a fifteenth-century overshot water-wheel for driving a pump. Note the crank, connecting rod and beam (bob). This would have been the type of engine used at Bere Alston mines in the 1470s. The drawing is from the medieval draughtsman Hausbuch and is dated 1480.

Miners were developing an understanding of practical geology – reading the ground – and this is illustrated by a court case, in 1472, involving miners who had sunk a shaft some 72 ft (22 m) deep before reaching a payable lode. The effort and cost of sinking through hard, barren rock to that depth shows that tinners' knowledge was sufficiently advanced for them to be able to predict with accuracy where the payable lode was likely to be. The local lawless gentry then attacked the miners with armed retainers and drove them away.[30]

The registration of tin bounds under the provisions of the ordinances of Prince Arthur in 1494 and 1496 gives us a fair picture of the state of the tin industry at the end of the 15th century. By then the western stannary of Penwith–Kerrier was dominant, and lode mines were increasing in number and overtaking stream works in importance. Groups of tin bounds, known as 'bals', were being exploited by large groups of miners in operations which were increasingly financed by local gentry and rich merchants. The cost of mining a lode was far higher than streaming alluvium and only the better-off could afford it. Trewellard Bal in St Just, Carnmeal Bal in Breage, Great Work and Trebullance Bal in Germoe, Carnkye Bal in Illogan, Polgooth Bal in St Ewe, and Poldice Bal in Gwennap were just some of the bals being worked by tin miners in the second half of the 15th century. Of 186 tin works registered in Penwith–Kerrier at the end of the century, 52 were mines and the rest stream works. Over a third of the tinners who registered were farmers working their own land, and there were no less than 36 entrepreneurs with shares or doles in three or more tin works. Of these latter, two held shares in five tin works, one in six, and

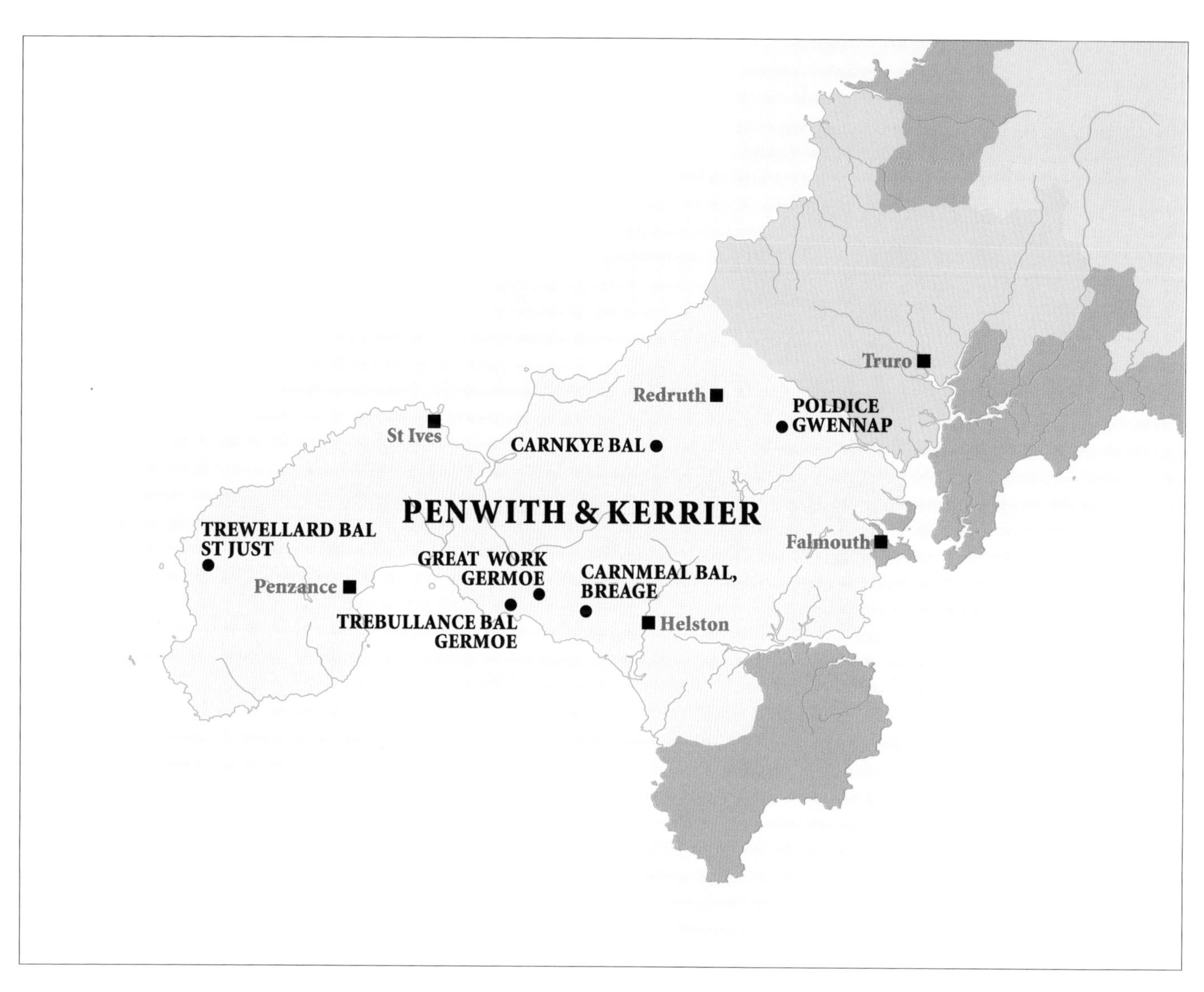

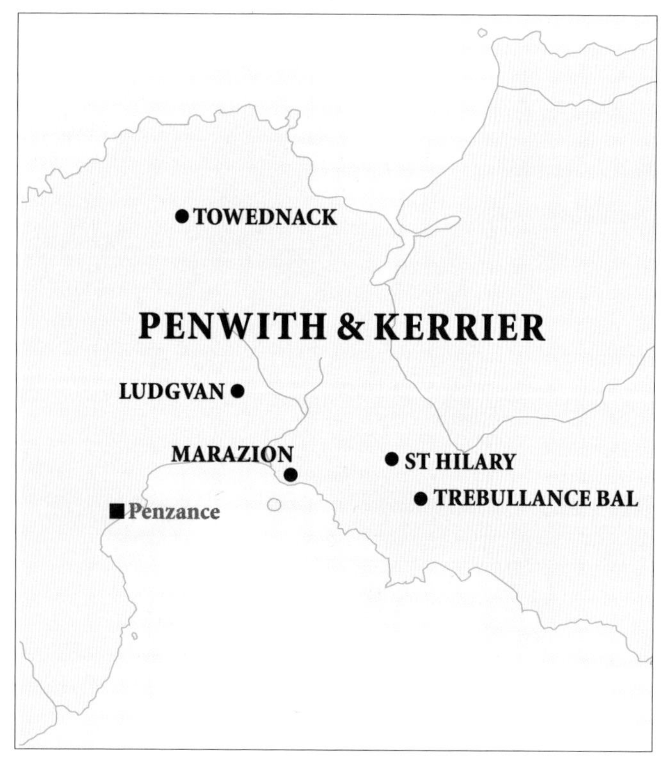

one – John Benyll (or Bevyll) – held shares in seven different tin works, which included a stream in Marazion, one in St Hilary, two in Ludgvan and two in Towednack, and a mine in Marazion. Trebullance Bal consisted of at least six lode mines with a large group of tinners, some of whom were involved in several of the small mines there. It was still being worked, under the name Lady Gwendoline Mine, in the late 1930s. The large openwork called An Coghan Bras (Cornish for 'The Great Openwork') was owned by Sir John and Sir Edmund Arundell, William Trevanyon, gentleman, and several other prominent local men at the end of the century, and it too was still being worked in 1914.[31]

The 1490s were a time of trouble in Cornwall, and with working costs increasing and control slipping away to the wealthier classes the average tinner was not happy. Henry VII (1485–1509) was a greedy monarch who took all he could from the people, rich and poor alike. Prince Arthur's ordinances imposed greater controls on the tinners as well as heavier duties on white tin. More officials, more red tape and a great deal more duty were a recipe for trouble, and in 1497, after demands for a heavy tax to pay for a war against Scotland, the tinners joined thousands of their countrymen in open rebellion. Starting in St Keverne, where the revolt was led by Michael Joseph an Goff, the rebels quickly moved on to Helston, the centre of the western stannary. Their numbers swollen by thousands of tin miners and streamers, they marched to Bodmin, where the motley army was joined by the lawyer, Thomas Flamank. As they moved towards London they were joined by thousands of others who shared the outrage at Henry's demands. Shaken to its foundations, the government called back the army, which was heading towards Scotland, and the two forces met at Blackheath. The issue was never in doubt, and the ill-equipped and poorly prepared Cornishmen and their allies were quickly cut down and dispersed. The leaders were executed. Many of those who escaped and returned to Cornwall joined Perkin Warbeck's rebellion later the same year and suffered again at the hands of the king's army.[32]

OPPOSITE ABOVE
Map of the principal tin 'bals' in the Penwith-Kerrier stannary in the second half of the 15th century. The cost of mining, as opposed to streaming, caused tinners to join together to finance the working of a number of tin bounds grouped in bals.

OPPOSITE BELOW
The spread of mines and streams where entrepreneur John Bevyll (Benyll) held doles or shares 500 years ago.

BELOW
The Fowey tin ingot is one of five found in 1898 in the wreck of a ship in Fowey Harbour. It is believed the ship sank in 1485.

The Charter of Pardon (1508)

There was another serious consequence of the tinners' rebelliousness and their perceived refusal to co-operate with the implementation of Arthur's ordinances. In 1508 Henry VII used the tinners' obstinacy as an excuse to withdraw Edward I's charter and replace it with his Charter of Pardon. Typically, Henry tacked on to the issuance of the charter a fine of £1,100, which was to be paid by the new rising class, the gentry, as they were then taking effective control of the tin industry. Thus, the situation that had obtained for the previous 130 years, whereby only those who did actual manual labour in the tin works could claim the privileges of being a tinner, was overturned in favour of those who had been expressly excluded by the definition agreed by Parliament in 1376 and later confirmed by Richard II. Not only did the Charter of Pardon not confirm the ancient rights of the tinners, it betrayed the very men the two earlier charters were designed to protect and encourage. Following its imposition upon the industry, the tinners were to witness an inexorable decline in their status as more and more moved from independence to being mere wage-earners and subject to the will of the ruling classes.[33]

CHAPTER THREE

ENTERPRISE IN THE STANNARIES

Tudor period (1508–1603)

OPPOSITE
Sixteenth-century shaft timbering in an illustration from Georgius Agricola's great work De Re Metallica *(1556) shows typical design for a two-compartment shaft. The main compartment was for hoisting ore, materials and water and the smaller one has a ladderway for man access. It is similar to shafts discovered in medieval English coal mines.*

BY CEMENTING THE DOMINANCE by the gentry of the tin industry, Henry VII's Charter of Pardon (1508) heralded a new confidence among the wealthier classes, creating a greater willingness to invest in enterprise and business. Capital investment was becoming more easily available for mining, as it was for other industries. International trade was changing and capital from Italy, Germany and France was financing the exploitation of metalliferous ore from all over the world, especially the Americas. By the middle of the century European bankers were paying for the flow of luxury goods from the East and gold and silver from Mexico and South America, thus altering the whole economic structure of Western Europe. By the end of the century the vast quantity of silver from the Americas accelerated the division of wealth that was beginning to appear in English society as elsewhere, and the inflationary effect of the increase in bullion flowing around Europe worked to the benefit of the better-off, often at the expense of those less able to protect themselves against such changing conditions. Cornwall was equally affected by these new economic trends, as was its tin industry, for surplus money created a demand for luxury goods, including pewter ware. Cornish tin refining was recognised as the most efficient in the world and it produced the finest tin metal. Pewterers from London and the Continent preferred Cornish tin to any other. New and bigger mines needed capital investment on an increasing scale, and the Cornish landed classes and merchant families were in a position to provide it. The Godolphins, Vyvyans, Beauchamps, St Aubyns, Boscawens, Edgecumbs and many other families were increasingly putting their money into mining – and being richly rewarded for it.[1]

The 'roughest and most mutinous men in England'

In contrast to this was the position of the poor tinner and wage-earning miner. Commodity price inflation was outstripping wage increases and the amount they were paid for their tin. During the 16th century many writers and officials commented on the poverty and status of these working tinners. Consequently, we probably know more of their situation and perceived character than for any other period until the

D
C
D
D

Sir Walter Raleigh, Lord Warden of the Stannaries between 1584 and 1603. Raleigh was a Devon gentleman and adventurer who became a favourite of Elizabeth I. He carried out various reforms to benefit impoverished miners.

19th century. Making allowance for a certain prejudice that may have coloured the views of the outside observer – due largely to the tinners' involvement in the revolts of 1497 and 1549 – we can still learn a lot about the men and their industry from these descriptions. Their truculent attitude was referred to by a government official in 1522, when miners were being recruited to serve in Henry VIII's army in France. Our 'faculty is to work underground, not above', they are reported as saying. All the more pertinent as it was Cornish miners who had successfully undermined the walls of Harfleur during Henry V's siege of the town in 1415. They were also considered dangerous, especially after the 1549 rising, and in 1586 Thomas Cely wrote to Sir William Cecil (Queen Elizabeth's chief minister) that Cornish miners were 'ten or twelve thousand of the roughest and most mutinous men in England'. In 1595 Sir Francis Godolphin, a magistrate and one of the wealthiest tin mine owners in Cornwall, commented on the extreme poverty and suffering of the miners, which he attributed partly to mines struggling with ever greater depths and technical difficulties. In 1597 Godolphin stated that there were '10,000 idle loiterers in this small county' due to a grave shortage of corn and other provisions, as well as the increasing cost of all the materials that were essential for deeper mining. In the 1580s Thomas Beare wrote that in the stream works of Blackmoor tinners might average £3 a year, whereas at about that time, according to Richard Carew, miners working underground received about 8d a day, or between £4 and £6 a year. In 1601 Sir Walter Raleigh, Lord Warden of the Stannaries, claimed that his use of pre-emption had stimulated employment and pushed up the wages of tinners from 2s to 4s a week. Hatcher has pointed out, however, that this still left the tinner poorer than a farm worker of the time.[2]

The figures for the number of tinners mentioned above by Cely and Godolphin appear to be high, for the actual number listed in the tinners' muster rolls for about 1535 is closer to 4000 than 10,000. Stoate, who transcribed and edited the lists, gave a total of 3457 tinners' names and estimated that, along with those listed elsewhere, the total number would have been about 4000. What is of interest is that well over half of those named tinners were in the stannary of Penwith-Kerrier, indicating that the shift to the west was complete by that time. The western parishes with the highest numbers were Redruth, Gwennap, Breage, Crowan, Lelant, Illogan, St Just and Camborne. John Norden, who carried out a place-by-place survey of Cornwall and drew maps of each hundred (the ancient county subdivisions) in the 1580s, gave a very sketchy picture of the extent of tin mining in Cornwall. He refers to tin production in a score of places west of Truro but, apart from Hingston Down, to the west of the River Tamar, he hardly mentions the tin industry elsewhere. Highly productive mining areas like St Neot and Polgooth were passed over almost without comment, despite the fact that St Neot alone had some 119 tinners listed in the muster roll. The stained-glass windows in St Neot Church are a witness to the success of tin mining on Goonzion Downs in the early 16th century.[3]

The stannaries and the crown

For more than half of the Tudor period the Duchy of Cornwall was dormant as there was no oldest son of the monarch, and from 1547 until 1603 it was continuously so. This left responsibility for the stannaries with the crown, and control was exercised through officials appointed by the government. For the most part the lord wardens were courtiers with west country connections, and for the last 19 years of Elizabeth's reign (1558–1603) Sir Walter Raleigh held the post. The vice wardens in Cornwall were men of significance in the tin industry, like William Beare, William Carnsewe and Sir Francis and Sir William Godolphin. One of the most contentious issues during those dormant periods was the use of pre-emption, and when men like Gilbert Brokehouse used their lease of pre-emption to exploit the vulnerable tinners, complaints were loud and insistent. Brokehouse had obtained his lease during the reign of Edward VI (1547–53) and exercised it oppressively into the reign of Mary (1553–58), when it took action by the privy council to right the wrongs and protect the oppressed tinners and hard-hit tin merchants.[4]

The Western Rising (1549)

It was during such a period of dormancy that the tinners of Cornwall became involved in the Western Rising of 1549, a rebellion that was spurred mainly by the rapid changes the government was bringing into the practice of religion. On the introduction of a new prayer book in English, a large number of men from Cornwall and Devon rose in armed rebellion and marched on Exeter. Of course religion was not the only complaint, for the rebels also demanded action on the severe food shortages and rising costs. Changes in agrarian practices, especially with respect to sheep, were also causes of anger, but for the Cornish there was also resentment at the imposition of the English language on their church services. It is ironic that it was Sir John Russell, until the previous year lord warden of the stannaries, who led the government's forces against the rebels. His foreign mercenaries were the ones who slaughtered the prisoners at Clyst when they feared a renewed attack by the men from Cornwall and Devon. It was the Lord Protector of England himself, Edward Seymour, Duke of Somerset, who sent Sir John to suppress the rebels, whom he had replaced as lord warden. Despite the frequently exaggerated claims about the number of rebels from Cornwall who were killed during and after the

The generall perambulation & delimatio of Cornwall
Milliaria sex
2 4 6
The Hundreds distin
A Penwith hund
B Kirrier hun
C Powder hun
D Pyder hund
E Weste hund
F Easte hund
PARS
MARE
BRI
Lezarde
lezard poy
The manacles
Pensans
Market Jew
Helston
Camburne
Redruth
Lanhern
Newlyn
Mounts Baye

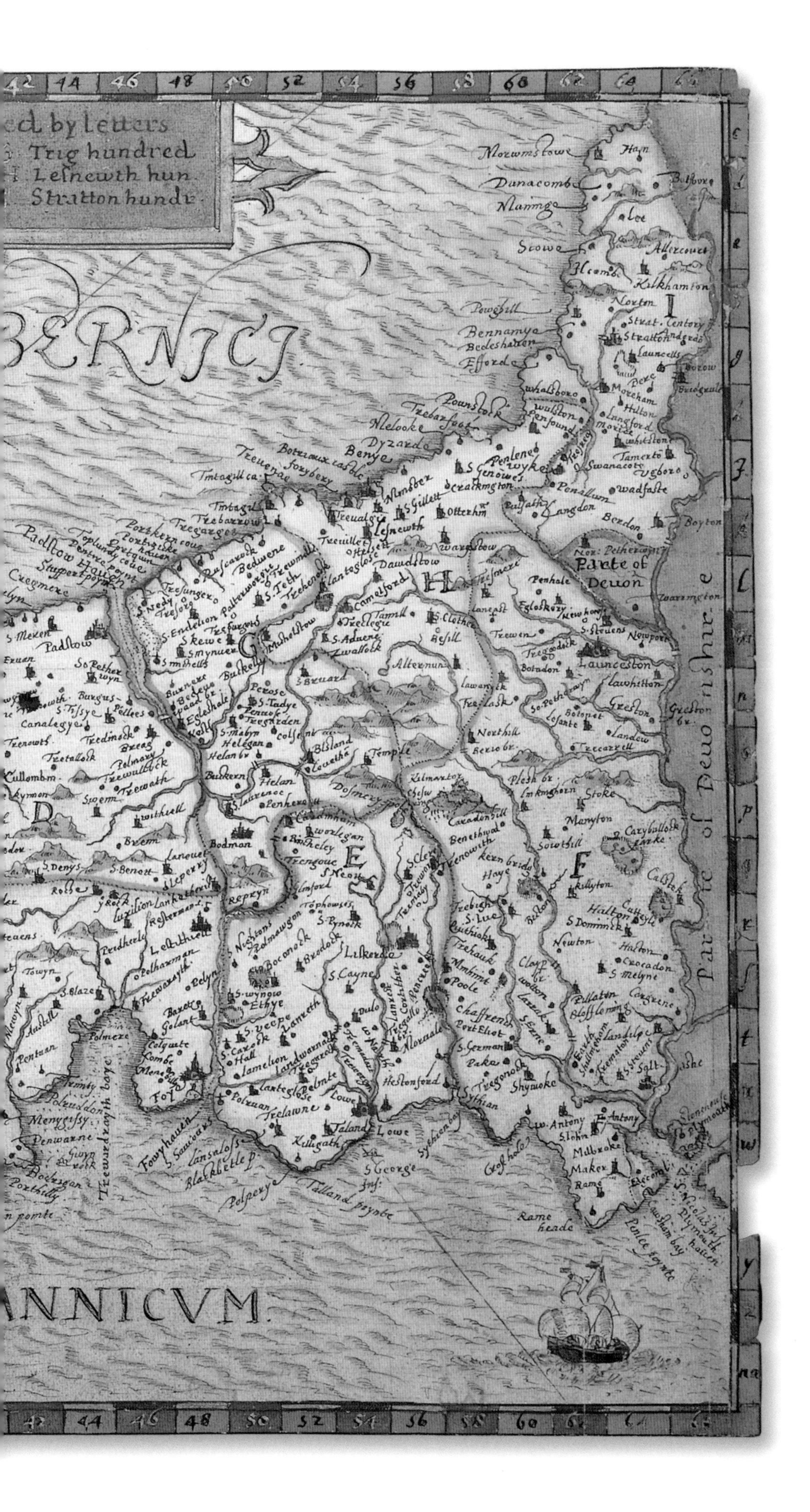

An accurate and detailed map of Cornwall drawn by John Norden in the 1580s. The map clearly shows towns, villages and waterways that were significant in the developing tin-mining industry. Norden also produced more detailed maps of the nine hundreds into which Cornwall was divided.

rising, the yearly output of tin, apart from 1549, does not appear to have been much interrupted by the losses of men. This indicates that far fewer ordinary Cornishmen were killed or seriously wounded during the troubles than has been claimed. The tonnage of white tin produced in 1547 was the highest for 150 years; the figure for 1549 was the lowest for half a century, but within five years it was back to its pre-rebellion level, where it remained for the next 13 years, when technical and economic factors caused its gradual decline. In purely numerical terms the working population in the stannaries appears to have been little affected by the rising or the events which followed the defeat of the rebels.[5]

THOMAS BEARE

Thomas Beare, who wrote the book *The Bailiff of Blackmoor*, was a man of unusual talents. Born between 1510 and 1520, he was appointed bailiff of the large stannary district of Blackmoor towards the end of the reign of Henry VIII (1509–47). It seems most probable that he was of a lesser gentry family which owned land and had interests in the tin industry in and around the parish of Warleggan. He was well read, quoting texts from Latin, Greek and French, and he wrote in good, clear English. He would have gone to one of the local grammar schools that were scattered around middle and east Cornwall before becoming involved with the tin trade. By the time he was 30 years old he had been appointed bailiff of Blackmoor, and he relates how he attended a coinage at Truro in that capacity in the last years of Henry's reign. By the time he wrote his account of the tin industry in 1586, he was experienced in every aspect of the tin works, the stannary court system, the weights and measures, the many and various water agreements and the hundreds of different tin bounds. He also knew most of the army of tinners who were involved in digging, dressing, transporting and blowing the tin. He settled disputes, negotiated new agreements, fined transgressors, supervised the movement of blocks of tin metal, ensured that the correct marks were put on them and generally saw that the whole of his stannary worked efficiently. He spoke affectionately of the tinners and clearly had the greatest respect for their honesty, diligence, skill, charity and humour. His account of the history of the tin industry and his knowledge of all the charters and legal decisions were almost certainly based on his access to the original stannary records, for which he was responsible and which, he said, were kept secure in Luxulyan Church tower.

A contemporary account of the traditional tinner

Thomas Beare's description of the tinners of Blackmoor in 1586 is most informative and probably typical of tinners throughout Cornwall. 'The most part of workers of black tyn and spaliers are very poor men as … [their] occupacon can never make them riche & chefly such tin workers … have no bargaines, but only trust to their wages.' After stating that they earned only about £3 a year, Beare added: 'yet must the worker find himself meat & drink, which is litle above 2d a day. This poor man happely hath a wife & 4 or 5 small children to care for which all depend upon his getting, whereas all his wages is not able to buy himselfe bread, then to passe over the poor mans howse rent, clothing for his poor wife & children besides diverse other charges dayly growing upon them. O God how can this poor man prosper?' Beare then describes the relationship between tinners, indicating why, despite their notorious poverty, outsiders often envied them. 'They are very charitable & mercifull toward the poor fellow workers, for at dinner tyme when they syt down together beside their tynwork in a litle lodge made up with turfes covered with straw & made about with hansome benches to sit upon, then every tynner bringeth forth out of their scrips or tynbags his victuals, his bread his bottle of drink as the riche tynners will lack none … Then is their charitie so great that if one two or 3 or els more poor men syt among them having neither bread, drink or other repast, there is not one among all the rest but will distribute of the largest sort with their poor worke fellows that have nothing, so that in the end this poor man … shall … be better furnished of bread, butter, cheese, beef, pork & bacon, than all the richest sort … [and] may carry the overplus toward the releife of his poor family home with him.'[6]

Another characteristic of the tinners was their outrageous sense of humour, which remained at the same pitch throughout the history of Cornish mining. Beare devoted several pages of his book to describing the 'merry devises comonly used among tynners', and also their unique names for various animals and birds. He calls it 'their mirth which they use without offence to any man'. Of the several hilarious examples Beare gives of this mirth, the most typical has to do with a man who inadvertently is foolish enough to tell his fellow to 'Kiss my arse!' Beare describes the reaction to such a suggestion: 'Then make they all a great showte, that it behoveth this fellow to have his arse cleane washed, that will have to be kissed, so to the river this fellow is borne with strength of the fellow workers, for it is vaine labour to strougle against them if he were never so tall a fellow. Adowne then ar his breches set & his arse, saving your reverence, well plonged in the deepest place of all ther is & insteed of a a towell to wipe his tayle, one cometh with scowpfull of gravell & litle rough garde. With this gard the poor man's tayle shal be so rubbed that the bloud shall follow & all this they say must needes be done to make his tayle cleane against that he shal be kissed.'[7]

The organisation of the isolated stream works had probably changed little over many centuries and Beare's account is very informative. After the Michaelmas coinage (assay), usually before 18 October, the tinners of each working would meet on their bounds and elect a captain, who had to be a reliable and skilled tinner with a share in the tin bounds. They would then take a square-cut rod and scratch on it the name of each shareholder in the tin works. This was used as a tally, which recorded every day missed by a dole holder. Those who missed work were fined 4d for every day they failed to turn up, and 6d if there were problems on the bounds, such as a flood brought on by torrential rain. If a man was sick he was spaled 2d a day, but if he was able to attend the workplace and merely observe he was not fined at all. This man would be well looked after by his fit mates, 'with no lesse care than if he were their father their brother or their naturall child'. Beare comments that 'I suppose that theis charitable orders among them geveth the poor men occasion to covet to be workefellowes among them'. If a wealthy shareholder could afford to send a servant to stand in for his spale, he was not fined. Some of them employed such spaliers on a regular basis, and these became skilled workers – but, unfortunately, still wage earners. At the end of each working period the captain called the tinners together and the books were balanced. The spales owed were calculated, other charges were added and the tinners all handed in their expenses incurred on behalf of the workings during the previous season's operations. One might have purchased a wheelbarrow, another some picks and another serge for a sieve, and so on. After fines were paid and expenses allowed, the profits were paid out to the individual tin bound owners. A year after renewing their bounds the tinners would be back in the itinerant stannary court, usually held every three weeks in a conveniently located inn or house, to renew their bounds. Some had been worked for centuries, and many retained their ancient names for generations. Beare was here describing the ancient, almost timeless, traditional way of working.[8]

Changing times – the trend towards larger, underground mines

The organisation of a large, well-financed tin mine stood in contrast to this unchanging, traditional system. It is true that many aspects of the long-established stannary system continued to be observed in the lode mines, but of necessity the organisation of the working itself had to be more modern and responsive to a capitalist system. In his *The Survey of Cornwall* (1602), Richard Carew gives us a clear description of how these mines were organised and how they worked. Unlike Beare, he was an outsider, and unlike Beare too, he had the gentry's patronising view of the tin miner. His admiration for them, when he showed it, was grudging. Beare, although probably of the same gentry class as Carew, was enthusiastic in his admiration for the tinner. Carew was shown around the tin mines of the Godolphins at

Breage and Germoe, and he was a keen observer with a good teacher – Sir Francis Godolphin himself. Great Work, the principal mine owned by the Godolphins, had employed some 300 men for generations when Carew visited it.[9]

Carew's account of sixteenth-century mining finance, organisation and practice is especially interesting. When a mine was found to be worth working, several gentlemen joined together in an adventure and put equal shares in to work it 'because the charge amounteth mostly very high for any one man's purse'. Some adventurers worked in the mine themselves and some employed 'hirelings', who were paid about 8d a day, or £4 to £6 a year. Mines, like bounds, had names by which they were known, and some names, like Wheal Fortune, were used in almost every mining parish. Great Wheal Fortune in Breage kept its name for over 400 years, whereas some changed as the owners did. Carew describes the role and responsibilities of the mine captain in terms that would be recognised in a Cornish tin mine at the end of the 20th century. He had to assign each miner to his task; he had to ensure that the mine was safe and well supported and give instructions to the binder or timberman; he had to ensure that appropriate timber was available; he had to place the pumps correctly and ensure that the workmen at surface and the miners underground had the correct tools. The miners' tools consisted of 'a pickaxe (poll pick) of iron about sixteen inches long, sharpened at the one end to peck, and flat-headed at the other to drive certain little iron wedges wherewith they cleave the rocks. They also have a broad shovel, the utter part of iron, the middle of timber, into which the staff is slopewise fastened.' He continued by describing the way a lode was worked, following it downwards and to the east and west as its tin values led. Where practical the lode was worked in an openwork from surface, but once it dipped at too much of an angle or went too deep the miners adopted the shaft and level method of following it. 'If the load lie slopewise, the tinners dig a convenient depth and then pass forward underground so far as the air will yield them breathing, which, as it beginneth to fail, they sink a shaft down thither from the top to admit a renewing vent, which notwithstanding, their work is most by candle light. In these passages they meet sometimes with very loose earth, sometimes with exceeding hard rocks, and sometimes with great streams of water.' These mines could be as deep as 50 fathoms or 300 ft (90 m), and the miners were hoisted up and down on a stirrup, with two men winding the rope on a windlass. The danger of bad ground and poor ventilation had to be coped with, and the rock could be so hard that even good miners might only advance a tunnel three feet in three weeks.[10]

Richard Carew, who wrote The Survey of Cornwall *(1602), was from a wealthy east Cornwall family of gentry. He visited the mines around Breage-Germoe in the 1580s with Sir Francis Godolphin, the greatest mine owner in Cornwall. His descriptions of the mines and miners are very perceptive and informative.*

It was the eternal problem of water that exercised the mine captains the most. 'For conveying away the water they pray in aid of sundry devices, as adits, pumps, and wheels driven by a stream and interchangeably filling and emptying two buckets … in sundry places they are driven to keep men and somewhere horses also, at work both day and night without ceasing.' But what seemed to impress Carew most was the way an adit was brought into a mine to act as a natural drainage tunnel. 'The adit they either fetch athwart the whole load, or right from the branch where they work, as the next valley ministereth fittest opportunity for soonest cutting into the hill … Surely the practice is cunning in device, costly in charge, and long in effecting … If you did see how aptly they cast the ground for conveying water, by compassings and turnings to shun such hills and valleys as let them by their too much height or lowness, you would wonder how so great skill could couch in so base a cabin as their (otherwise) thick-clouded brains.' In this somewhat backhanded way the educated Carew acknowledged the undoubted intelligence of these unschooled mine captains.[11]

Admiral Edward Boscawen,Viscount Falmouth (1711–61) was a valiant sea captain who defeated the French in spectacular fashion and was known as 'Old Dreadnought' by his sailors. The Boscawens were involved in tin and copper mining as mineral owners and adventurers from the 16th century, and remained involved until the closure of Wheal Jane Mine in 1991.

Carew described the Cornish stannary system with its four districts and ancient coinage towns. The officers were the same as they had been for centuries. The rules about bounding remained the same, with tinners allowed to bound without restriction on unenclosed waste ground but needing the land owner's permission on enclosed land which had not formerly been bounded. Bounds were marked by the simple device of raising a pile of three turves at each corner, and they had to be renewed annually. The lord of the soil had the right to place his own workmen in a tin work up to a one-fifteenth share.[12]

Both John Norden (in his *Speculum Magnae Britanniae pars Cornwall*, c. 1610) and Carew described the process of crushing, grinding and concentrating the tin ore, but Carew also noted the improvements that were being made and new techniques that were in development. The larger lumps of rock were broken by men with sledgehammers, after which the ore was carried to the water-powered stamps, consisting of up to six iron-shod stamp heads, which pounded the dry ore to sand and fines. By the middle of the 16th century, however, dry stamping – which had been followed by heating the ore to remove any moisture prior to grinding with a crazing mill – was being superseded by wet stamping, which gave a finer and more easily washed pulp. The crazing mill was then removed to the end of the process, where only the 'crust of the tails' was ground. Similar improvements were being made to the smelting process, and Carew specifically mentions a 'Dutch mineral man' who advised Sir Francis Godolphin on a more efficient design of his blowing-house. He also brought new ideas on treating hitherto rejected tailings.[13]

The 1590s – an industry in crisis

As the 16th century drew to an end the tin industry once more found itself in crisis. Annual output dropped by over a third between 1592 and 1598 and the great depth of the mines, increasing water inflow, problems of poor ventilation and the cost of materials all hindered the industry's profitability. The working tinners and miners were becoming poorer. Wage and commodity inflation, population growth, rent increases, pressure on crop-growing land, the effects of the new poor law and vagrancy acts, together with the difficulty of finding steady and reliable employment, all made the struggle for survival of the poor even more difficult. And if the wage-earning miner was struggling, the independent tinner was in no less difficulty. Every contemporary Elizabethan writer on the subject of Cornish tin makes the same point: the usurer, money man and merchant were taking advantage of the poverty of the tinner, buying his tin cheaply and selling it at a greatly increased price to the London pewterers.[14]

Faced with this severe depression in the industry, the government decided to intervene. The crown's income from tin duty was threatened, the gentry (who were the crown's principal supporters in the stannaries) were losing money and threatening to channel their capital into agriculture, and the ever-present poor were becoming a potential social problem. The last thing a Tudor government wanted was to upset 10,000 of the 'most mutinous men in England'. Raleigh, as lord warden of the stannaries, led the government's attack on the problem in the time-honoured way – by the use of monopoly. Pre-emption was used to guarantee a higher price for tin and a living wage for the tin worker. The tin price was to be linked to inflation for a period of years, and this, it was hoped, would solve the problem for all concerned. In 1601 Raleigh claimed that his measures had doubled the weekly wage of the miners and production was certainly maintained at a reasonable level, although many of the underlying problems remained, especially for the poor.[15]

Meeting the challenge of wet mines

With the mines struggling against worsening technical problems, there was an upsurge in creative and innovative thinking with respect to the main problem of mining: dealing efficiently and economically with water drainage. In the 1550s a German university lecturer and engineer, Georgius Agricola, described and illustrated over twenty devices for dewatering mines. These pumps were powered by men's muscles, horses or water. Overshot and undershot water-wheels were used to power suction, forcer, bucket, scoop and rag-and-chain pumps. Exaggerated claims were made for these inventions and for the most part their standards of efficiency and economy can hardly have made them practical solutions to the problem of mine drainage. Patents were taken out for mine pumps throughout the last quarter of the century, with Englishmen and Germans vying with each other to offer solutions to this expensive problem. In the 1570s alone three Englishmen and a German applied for such patents. Although none of these pumps answered the problem completely, they were typical of the energy that was applied at the time to solving the technical problems of mining and – necessity being the mother of invention – a plethora of inventions was tried out.[16]

Education, literacy and the miner

One reason for the increasing inventiveness among the intelligent miner and his middle-class employer was the enhanced opportunity to gain a basic education. More families were sending their sons to the increasing number of local schools. At Glasney College, Penryn, in the 1540s there was a school for teaching poor boys their letters, and at Week St Mary there was a school which was transferred to Launceston after the Reformation. At both Liskeard and Bodmin reformation commissioners sought out places to establish schools for local boys, and these were followed by others at Saltash and Truro. More of the sons of the wealthy were going to Oxford or the Inns of Court in London, more of the sons of skilled workers were attending local schools, and, as education and literacy filtered down, more of the working classes were able to read and write. Particularly during the second half of the 16th century there was a trend to send sons to the Inns of Court rather than to university. The

Two illustrations from Agricola's De Re Metallica *of 1556. The one on the left shows early dry stamps operated by an overshot water-wheel, and that on the right shows pumping machinery powered by two overshot water-wheels, one at surface and the other underground, just above the adit drainage level. This type of stamping mill was used in Cornwall in the late 15th century, and eventually pumps of the type illustrated were also introduced.*

education offered there was of a more practical kind, and with estate and business management high on the agenda of parents it proved a sensible choice. John Chynoweth estimates that between 1553 and 1603 more than three times the number of Cornish boys attended the Inns of Court than attended university.[17]

This trend was not just a Cornish phenomenon for – as the modern historian Christopher Hill has shown – throughout England at that time, as in no other country, ordinary men were learning to read practical, intellectual and scientific books. This trend towards universal literacy was encouraged by the fact that more and more people were now able to read the Bible regularly in the vernacular and understand it. Hill writes: 'In sixteenth-century England there was a greedy demand for scientific information. Thus Thomas Langley in 1546 translated Polydore Virgil's *De Inventoribus Rerum* "to the end that also artificers and others, persons not expert in Latin, might gather knowledge … by reading thereof".' The result was that even shepherds, like Robert Wyllyams, 'keeping sheep upon Seynbury Hill', Gloucestershire, were willing to pay 14 pence for a copy (recall that an underground miner's wage in the 1580s was 8d a day). Hill saw this phenomenon as 'something like an adult education movement', with intelligent men throughout the country seeking to learn more and to use their newly acquired knowledge to improve their lives, their working conditions and their working practices. Innovation and invention are the natural children of such a movement, and Cornish miners were at the very forefront of it. Thus, despite the crisis facing the tin industry, there was also a feeling of confidence and optimism about. Nowhere was that confidence more apparent than among the hard rock miners themselves. Poor they might be, but as William Carnsewe wrote to Thomas Smythe of the Mines Royal in 1584, Cornish miners were 'as skillfull in mining, as hard and diligent labourers and as good cheap workmen in that kind of travail as are to be found in Europe'. Carnsewe even issued a challenge that Cornish miners be given a mine and the Germans be given another, and he reckoned that the Cornishmen would work faster and more efficiently than the foreigners. As the Tudor dynasty ended Cornish mining was moving into a new era in which innovation and ingenuity were to be as important as traditional methods of working, which had changed but slowly over the preceding thousand years.[18]

Another drawing from Agricola's sixteenth-century work describes a typical mining scene. Shafts, windlasses, hand-pushed trucks for ore and miners working underground are all depicted here.

CHAPTER FOUR

CROWN, PARLIAMENT AND THE GLORIOUS REVOLUTION

17th century (1603–1690)

OPPOSITE
An engraving made in 1842 of large openwork at Carclaze, near St Austell, illustrates the wide use of water power for operating pumps and stamps, and for washing the cassiterite in buddles. A bal maid is seen attending the buddle. The scene is typical of such tin works throughout the 17th, 18th and 19th centuries.

THE 17TH CENTURY witnessed fundamental and radical changes in several important aspects of Cornish life. Maritime trade expanded, new sources of capital were available, global trading links were established, wide-ranging constitutional changes were made, tighter central government control of the regions was secured and new financial institutions like the Bank of England and the National Debt came into being. The long-running and bitter struggles between Parliament and the Crown, which started almost with the arrival of James I in 1603 and ended with the so-called Glorious Revolution in 1688–89, totally changed the political and constitutional landscape. Cornwall was involved in these struggles at every level, with her members of Parliament sometimes leading the battles on both sides of the dispute. The social, demographic and economic effects of this century of conflict and warfare on the Cornish tinners and their industry were profound.

Developments in the early Stuart period

The Stuart era began as the Tudor period ended, with poverty among the tinners widespread and confidence in their own skill and mining expertise growing. And to an increasing degree this self-belief was coming to be shared by wealthy and powerful industrial entrepreneurs in other parts of the country. In 1608, when the new king, James I (1603–25), wanted skilled miners to develop the Scottish silver mines, he bypassed the German miners employed by the Mines Royal at Keswick and brought in Cornishmen despite the much greater distance they would have to come from. 'The miners of our Dutchie of Cornwall … are held to be the most experienced and most exercised in such woorks of all other our people.' The deputy warden was instructed to send 20 experienced tin miners to Scotland, accompanied by 13 men with specific skills, including two captain timberers, two smiths, two 'ruffbudlers' and seven others with skills in aspects of concentrating and smelting ore.[1]

At the level of mine engineer the Cornish were also recognised as leading the way, for in the 1620s, as the Crown was trying to develop the silver–lead mines of Wales,

it was a Cornish mining engineer, Sir Francis Godolphin, who was sent for. At the behest of the Privy Council he travelled to Cardiganshire to inspect the mines and evaluate and assay the ore. He was described as 'an able Cornish mining engineer' who not only understood tin ore, from his home parish of Breage, but was also a skilled assayer of lead and silver ores. Godolphin was not alone, for there was a growing number of such Cornishmen who were learning and developing the skills necessary to open up and operate the increasingly sophisticated mining operations in Cornwall and elsewhere. Another man who made his name at that time was Hannibal Vyvyan (Vivian) of Camborne. In January 1634 he registered a patent for an engine which pumped water from deep mines as well as hoisted ore and other materials up mine shafts to the surface. The original wording of the patent says: 'Hannibal Vyvian, hath after much paynes and expences, invented and brought to perfection an Engine of great use and profit … whereby one man alone can easily drawe as much water as otherwise four, and otherwise hoyse (hoist) all weighty thinges out of the saide Tin well.'

In the quarter of a century before the Civil War and in response to mine drainage problems, 23 pumping engines or modifications to such engines were patented in England. All the patentees sought a remedy for the most difficult problem facing the constantly deepening metal mines of the country – draining the water that came into the workings – and several specifically mentioned tin mines as their area of concern. Some other practical improvements were carried out by working men who altered and modified their machines to work more efficiently. In about 1640 a young tinner called John Tomes (Thomas) invented a method of automatically cutting off the water flow to the wheel of the stamp mill when the coffer was empty. As the coffer held the ore that was being broken by the stamp, this protected it and the mortar stone beneath from damage by the action of the heavy stamp head when the stamped ore had passed through.[2]

Elizabethan attempts to improve the tin miners' lot were continued during the first decades of the Stuart era. The policy of leasing tin pre-emption to guarantee a better price for tin and a living wage for the tinner was generally successful, and the price was usually fixed for periods of seven years. As a result, tinners were no longer dependent on tin buyers for loans. The scheme was continued with varying success until the 1640s, when pre-emption was abolished under the Commonwealth government of Cromwell (1649–60). However, when inflation overtook the price of tin, hardship usually followed. The poorer miners tended to support this use of pre-emption because it brought some stability, but the wealthier dealers were generally opposed, asking for the right to buy and sell the tin to whom they wished. G.R. Lewis has pointed out that the dealers loaned money to the tinners at a 20 per cent rate of interest and bought the tin cheaply. The Elizabethan government had sought to remedy this situation by supplementing price control with a system of cheap loans to tinners. In 1577 Sir George Cary had assumed the tin monopoly and loaned £4000 to the tinners at a low interest rate. Eventually, both the lessee and the Crown made interest-free loans to enable the tinners to survive until they sold their tin. By the middle of the 1620s, however, this provision had withered, and in 1624 the Tinners' Parliament, established through the provisions of the 1508 Charter of Pardon, petitioned the king for a £4000 loan on 'reasonable security' to help the poor tinners. As commodity price inflation continued, the gap between earnings and the cost of living was again crippling the tinners, and in 1636 they again petitioned the king for help. Mine owners were losing money, the poor were getting desperate and mines were being abandoned as unprofitable. More seriously for the collectors of tin duty, the income from tin dropped by a quarter between 1625 and 1638. Unfortunately, as this was a time when Crown officers were squeezing every penny they could from all and sundry, when the country was enduring the so-called 'eleven years tyranny', or the personal rule of Charles I (1625–49), when Parliament was suspended and the nation was in political turmoil, the particular problems of the poor in the extreme south-west of the country hardly merited much attention from the government.[3]

THE STANNARY PARLIAMENT OR CONVOCATION

The Stannary Parliament or Convocation had its origins in the provisions of the Charter of Pardon of 1508. It was a unique institution in that it was not only a parliament which represented a single industry but, theoretically, its powers paralleled those of the Parliament at Westminster. It could veto legislation from Westminster if it affected the tin industry, although there are no records of it ever doing so. The Parliament consisted of six 'stannators' from each stannary district, and these were appointed by the mayor and corporation of the towns of Launceston (Foweymore), Lostwithiel (Blackmoor), Truro (Tywarnhaile) and Helston (Penwith and Kerrier). The full legislative Parliament met once in the 16th century (in 1588), three times in the 17th century (1624, 1636 and 1686–88) and twice in the 18th century (1750 and 1752–53). It also met a few times to settle questions of pre-emption. The stannators always came from the wealthy classes, with baronets, gentlemen and rich merchants the only names on the lists – there were no working tinners among the stannators!

The Civil War

For the tinners it was the outcome of the Civil War (1642–51) that was to change their situation. The decade between 1650 and 1660 saw the Commonwealth government adopt a policy of non-interference. Coinage duty and pre-emption were abolished, thus freeing the tinners to sell to whom they wished at the price they could get. Abolition of the tin-buying monopoly and the law of supply and demand brought a short-term increase in the price paid to tinners and mine owners. The Tinners' Grievance stated: 'The preemption being resigned by the farmers, the price of tin rose. Multitudes of tradesmen left their callings for that of mining. Still the prices rose. Old, abandoned works were filled again and new ones taken.' Under the Commonwealth the price of tin went up from £60 to £120 a ton.[4]

By the end of the 1640s the victorious Parliamentarians were short of cash and in order to pay their troops it became necessary to raise large amounts of money quickly. Church land and the lands of royalists were sold, but Crown estates were a special target, and the Duchy of Cornwall was high on the list for confiscation and sale. In 1649, Parliament commissioned a survey of the property, which detailed every tenement on the several manors that made up the Duchy. The Duchy's holdings were split up and sold to local gentry and parliamentary soldiers. Although some attempts were made to change the manorial customs under which the tenants held their land, things continued much as before. Among other things, the survey gave some indication of the extent of tin working on Duchy land, with blowing-houses, stamp mills and crasing mills referred to in many locations between Wendron in the west and Calstock in the east. The manor of Helston, in Wendron parish, contained nine blowing-houses, eight stamp or knocking mills, two scoffe mills and a crasing mill. Following the death of Oliver Cromwell in 1658 and the return of Charles II a little less than two years later, these lands reverted to the Crown and the Duchy was re-established.[5]

The Restoration

The decade of prosperity ended with the Restoration in 1660. During the 1650s the wages of skilled miners had been 30s a month, but thereafter they gradually dropped. In 1667 a skilled miner earned 28s, an ordinary tinner 20s, and an unskilled labourer 16s a month. Within a short time a tinner's wage was down to 14s a month. Paradoxically, as wages dropped the mines produced more, and the output of tin went up from 911 tons in 1667 to 1399 tons in 1683 and 1407 tons in 1700 – and all this despite the closing of mines and many miners being thrown out of work. Those who could find work were forced to work long hours on both day and night shifts. The last decades of the 17th century saw miners throughout the stannaries begging for food and rioting when supplies were short. Their diet became one of potatoes, barley-thickened gruel and barley bread, with meat scarce and of poor quality. In 1690 a large mob of miners descended on a ship in Falmouth harbour and looted the cargo of salt. Attacks on food stocks became common as desperate men sought to feed themselves and their families. By the end of the century miners were seeking the remedy that had been used by Elizabeth's officials a century earlier: a guaranteed tin price and wage support through the device of of pre-emption. In 1703 an enormous gathering of tin miners surrounded the stannary convocation hall in Truro. Their demand was that the gentry who controlled the tin industry through the Tinners' Parliament accept the Crown's offer of pre-emption. After the Restoration the merchants and landed gentry who made up the convocation had gradually reimposed a new monopoly on the purchase and sale of tin, so that once again the poor tinners and wage-earning miners were at their mercy. The hungry miners saw relief in pre-emption as it could at least bring back some control on the price of tin.[6]

If the tinners' situation hardly improved following the Restoration, the industry itself saw many changes. The introduction of new inventions and innovations in all aspects of the tin industry was a feature of the time. Between 1660 and 1700 33 patents were registered for new mine pumping engines or improvements to existing ones. Half-a-dozen of these inventions were by Cornishmen, among whom Henry Vincent, Charles Vivian, John Tredenham and John Trewren made significant contributions. Most of these pumping engines were water-powered, but at least one, invented by John Johnson in 1674, was powered by a windmill. Such an engine was erected at Relistian mine, in Gwinear, and another, in the late 17th century, in Wendron, on 'a great tin work' at Tremenhere. In 1687 Sir Robert Gordon invented a pump which used a plunger pole to force water up a pipe and had non-return valves to enable the water to rise in stages. Richard Bullock also made significant improvements to mine pumps in 1690. Over a dozen patents were registered during that period to improve smelting techniques, and others were concerned with more efficient ore crushing and stamping machines. Engines like the horse-whim, for hoisting ore up deep mine shafts, were also being introduced throughout the mining districts.[7]

By the 1690s water-powered pumping engines had become quite sophisticated and sufficiently powerful to drain mines at the depths which the deeper ones had reached. In 1778 William Pryce described the pumping machinery in existence in the last decade of the 17th century: 'About four score years back, small wheels of twelve or fifteen feet diameter, were thought the best machinery for draining mines; and if one or two were insufficient, more were often applied to that purpose, all worked by the same stream of water. I have heard of seven in one mine worked over each other. This power must have been attended with a complication of accidents

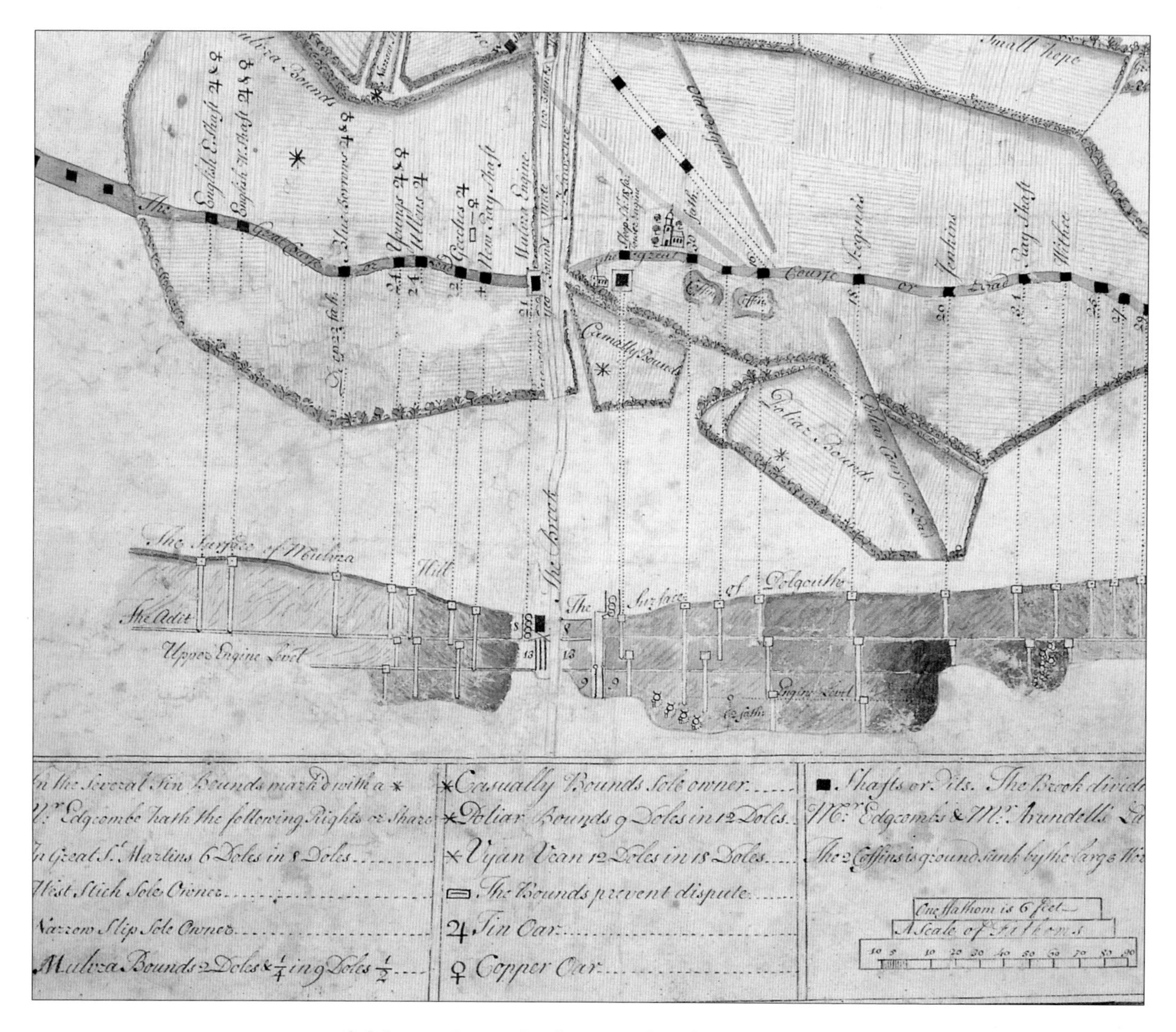

This plan and section of Polgooth Mine in 1695 is possibly the earliest extant long-section of a Cornish mine. It shows the levels, pumping arrangements and adit system. The whole surface layout is shown on the plan.

and delays.’ These wheels were placed in shafts one below the other, with a stream of water from a launder directed over each wheel in turn. Each wheel had a crank with a sweep rod attached, which in turn lifted a rod that extended down the shaft to pump the water. This series of wheels extended down to adit (drainage) level, through which the water from the wheels ran out of the mine. In 1699 this type of engine was erected at Carnkye Bal, an ancient group of tin mines in Illogan parish. A variation on this engine was called a ‘tower engine’, and it worked in precisely the same way as those sited in shafts, except that the wheels extended down through a wooden tower that stood over the shaft. There was such an engine on Poldice mine and another at Polgooth. A plan dating from the 1690s shows the Polgooth tower engine and another of the same type underground on the same mine. As Pryce commented, these engines were complicated and difficult to maintain, but they did drain the mines to the depth required at that time, which for some of the deeper mines was approaching 100 fathoms or 600 ft (180 m).[8]

An anonymous description of tin mining in 1671 shows that technically some aspects of the industry had hardly progressed since the time of Richard Carew. Ground-breaking methods had remained fairly static, as had ideas on finding and developing new sources of tin. Although there had been improvements in ore crushing, processing and smelting, costean pits or assay hatches remained the principal way of discovering a lode, and most of the machinery on surface was still primitive, if effective. With picks and plugs-and-feathers continuing to be the usual method of breaking the rock, adits could only be driven over fairly short distances and the amount of ore the miners could break in a shift remained modest. The method of opening up a tin lode, once it was discovered, by sinking a series of exploration pits, or 'assay hatches', was ancient. The assay hatch where the lode was discovered became an access shaft from the bottom of which the lode was followed downwards by 'shambles', or large steps. Each step was about 6 or 7 ft (2 m) high, so that broken ore could easily be thrown up to the next step by 'shovelmen'. When the lode became payable, drifts or levels were driven on its course to the east and west of the shaft. The levels were about 3 ft (1 m) wide by about 7 ft (2 m) high. Normally these tunnels were driven every five fathoms or 30 ft (9 m), leaving a convenient block of ground to be worked ('stoped') by the miners. Once the shaft was down to adit level, water was raised to it by buckets, 'leathern bags' and pumps. Most of these latter were primitive 'rag-and-chain' pumps, and were operated by men at adit level who wound the continuous chain up the pipe from the sump below. Balls of rag attached to the chain pushed the water up to adit level. These rarely worked efficiently over heights greater than 20 ft (6 m). The account says that drifts were advanced by three 'beelemen' ripping away the ore and deads (waste) while two 'shovelmen' removed

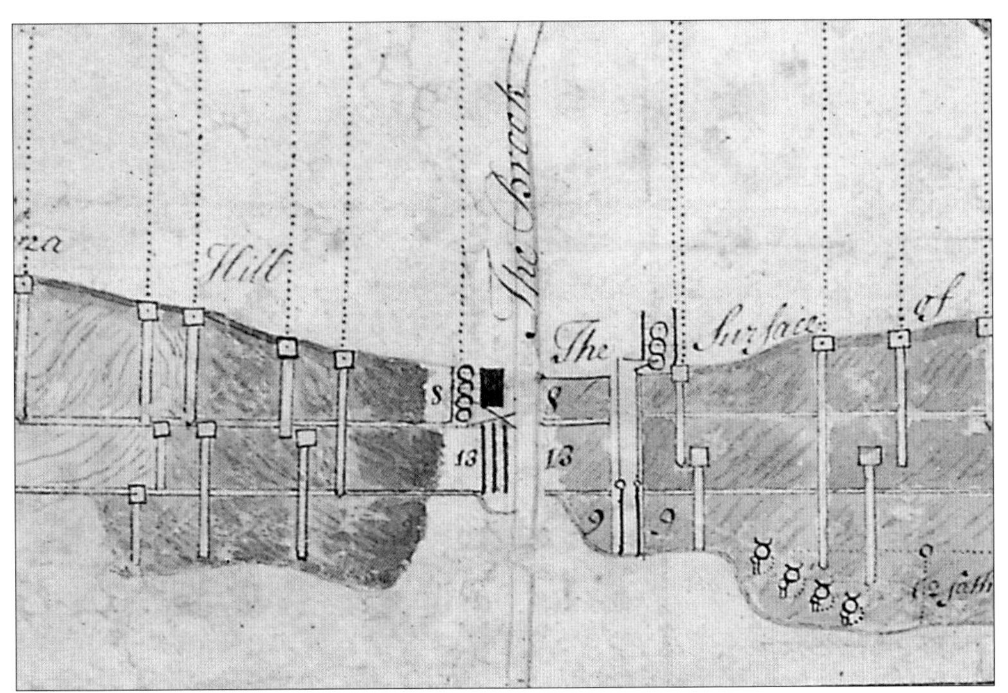

In this detail from the 1695 section of Polgooth Mine the two water-powered pumping engines are clearly illustrated, one with the wheels over each other in the shaft and the other, called a tower engine, with the wheels in vertical line in a wooden tower. There was another such tower engine at Poldice Mine, Gwennap, at the same time. The miners are shown working at the bottom of the mine on large benches.

it. They would barrow it back to the shamble, to be thrown up in stages to surface, or to the shaft to be hoisted up with a 'winder with two keebles' (kibbles). This description, contributed to the 1671 issue of the Royal Society's *Philosophical Transactions*, was written by a man who lived close to Hingston Down and was familiar with the mines on the Devon border and on Dartmoor. The mines in that area were not as advanced and well organised as the larger mines in the west of Cornwall, so the practices described are not necessarily typical of the most productive mining districts.[9]

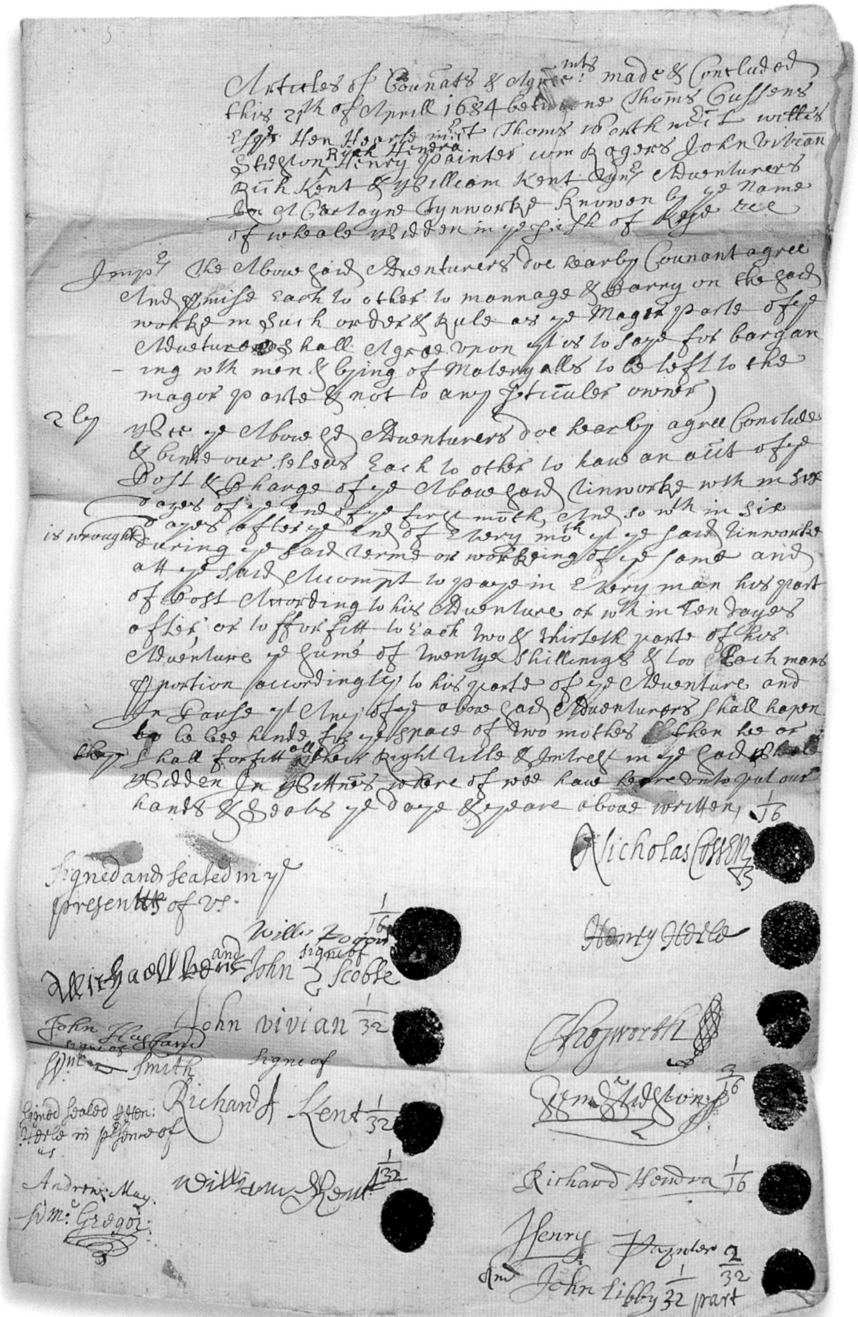

These are articles of agreement for a Cost Book Company drawn up between ten adventurers in Wheal Widden tin mine at Baldhu in Kea Parish in 1684. It is possibly the earliest extant document of its kind.

A mining revolution – gunpowder blasting

A major advance in Cornish mining technology came with the introduction of shothole blasting with gunpowder. The difficulty with gunpowder had always been how fire could be got to the explosive charge safely. Experiments in Mexican silver mines in the 16th century had failed, but in the 1620s a Hungarian miner, Caspar Weindl, managed to drill a hole, charge it with gunpowder, stem it with a wooden plug and fire it successfully. The flame travelled to the charge along a groove cut along the side of the wooden plug. This groove Weindl had rubbed thoroughly with gunpowder before the plug was driven into the hole. A trail of powder was laid to the plug and lit by a candle. From Hungary the method was taken to Germany and thence to the north of England. Eventually it was introduced into the lead mines of the Mendips, before being taken on to Cornwall from Somerset by Thomas Epsley in June 1689. It seems that the Godolphins were responsible for bringing Epsley to Cornwall, and it was their mines at Breage and Germoe that were to be the first beneficiaries of this huge advance in mining methods. No invention was to transform mining as rapidly as did the introduction of shothole blasting with gunpowder. Within a decade the method was in use throughout west Cornwall, and the results were spectacular. Data from that time on the rate at which tunnels were driven show that there was a dramatic increase in the daily footage achieved by miners. Levels were often extended by more than a foot in one shift. One hundred years earlier Carew had written that it often took a week for levels to be driven that far. Drainage adits – which had sometimes taken generations to bring home to mines – were being driven in a fraction of that time. The tonnage of ground broken by miners working in underground stopes rose significantly as they learned to use gunpowder, and, by the early years of the 18th century, gunpowder was being used for all types of mine work: sinking shafts, making raises, stoping and driving levels.[10]

Mine ownership during the 17th century

During the 17th century the move towards control by the gentry of all aspects of tin production advanced considerably, such that even tin-streaming bounds came increasingly into their ownership. Estates employed 'tollers', who ensured that whole groups of tin bounds were obtained and then annually renewed, resulting in exclusion from whole districts of the small, independent tinner. Thus, in 1639 Thomasyn Coryn, a gentleman's widow, held 15 pairs of tin bounds in Kenwyn parish, which were renewed yearly on 5 June; in 1676, in Wendron, the Pendarves family held 25 setts of tin bounds, which they sold to their relatives the Bassets, who renewed them throughout the 18th and 19th centuries; in 1684 John St Aubyn left 14 pairs of tin bounds between St Hilary and North Downs; in 1690 Edward Vincent sold his tin bounds in St Agnes, Kenwyn, Kea, Redruth, Gwennap and Perransands to John Williams of Probus. All these tin bound owners were wealthy members of merchant or land-owning families. Control by the gentry of the Convocation (Parliament) of Tinners was also increasingly evident as the century progressed, and after the Restoration convocations called in 1674 and 1687 opposed the reintroduction of pre-emption as they believed it restricted their ability to profit from the sale of their tin. The ordinary tinners tended to be in favour of its reintroduction because they believed that, although imperfect, the system did bring a measure of stability to the price of tin and, thereby, to their wages.[11]

Towards the end of the century some mines were employing very large workforces indeed, and overall the numbers of miners had increased. One thousand men worked on St Agnes Bal, Poldice mine employed between 800 and 1000 men and boys, and Celia Fiennes, who rode through Cornwall in 1696, estimated that about 1000 men were employed in the mines to the west of St Austell, including Polgooth. Unfortunately, estimates of the total number employed in the tin industry vary greatly. One estimate gives a figure of 60,000 men, women and children working in the mines and streams in 1681, another source suggests 20,000 as the number employed in 1700, and yet another states that some 8000 men were employed at that time. James Whetter has produced statistics which show that in about 1600 less than 3 per cent of Cornishmen were employed in the tin industry, whereas by 1700 this had risen to 8 per cent. What we do know is that many mines were very large and employed great numbers of workers, and among them were mines in Wendron, Gwennap, St Agnes, St Just, Camborne, Gwinear, Breage, Germoe, St Hilary, Warleggon, St Neot, Illogan and St Mewan. Polgooth was expanding, Poldice stretched across a mile of downland, Great Wheal Fortune and Relistian were enormous, and Tolcarne and Great Work were still producing considerable quantities of tin. An idea of how many men were working on production in some of these mines can be gauged from the number killed in just one stope at Relistian, in Gwinear, in 1681, where a rock fall caused by heavy rain buried 24 miners.[12]

A lease agreement granted by William Beauchamp of Trevince in Gwennap in 1699 for a streaming sett in Gwennap, which included watercourses, buddles and tin stamps, known as 'Trefins Higher Stamps'.

Developments in the business and legal setting

Following the revolution of 1688–89, when Parliament took effective control of the country, new conditions were created whereby business and industry could grow, and global trade – already beginning to dominate the economy of the country –

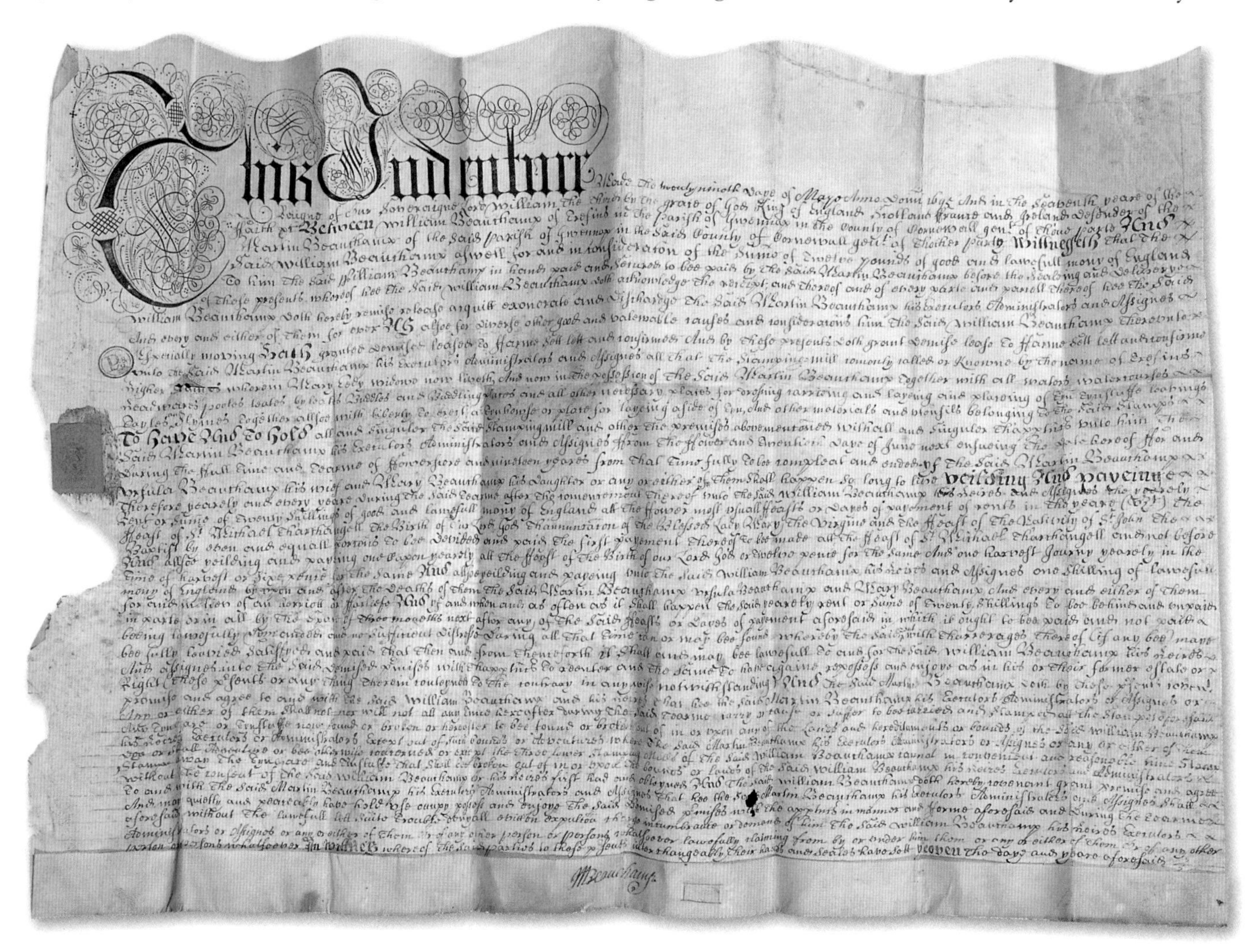

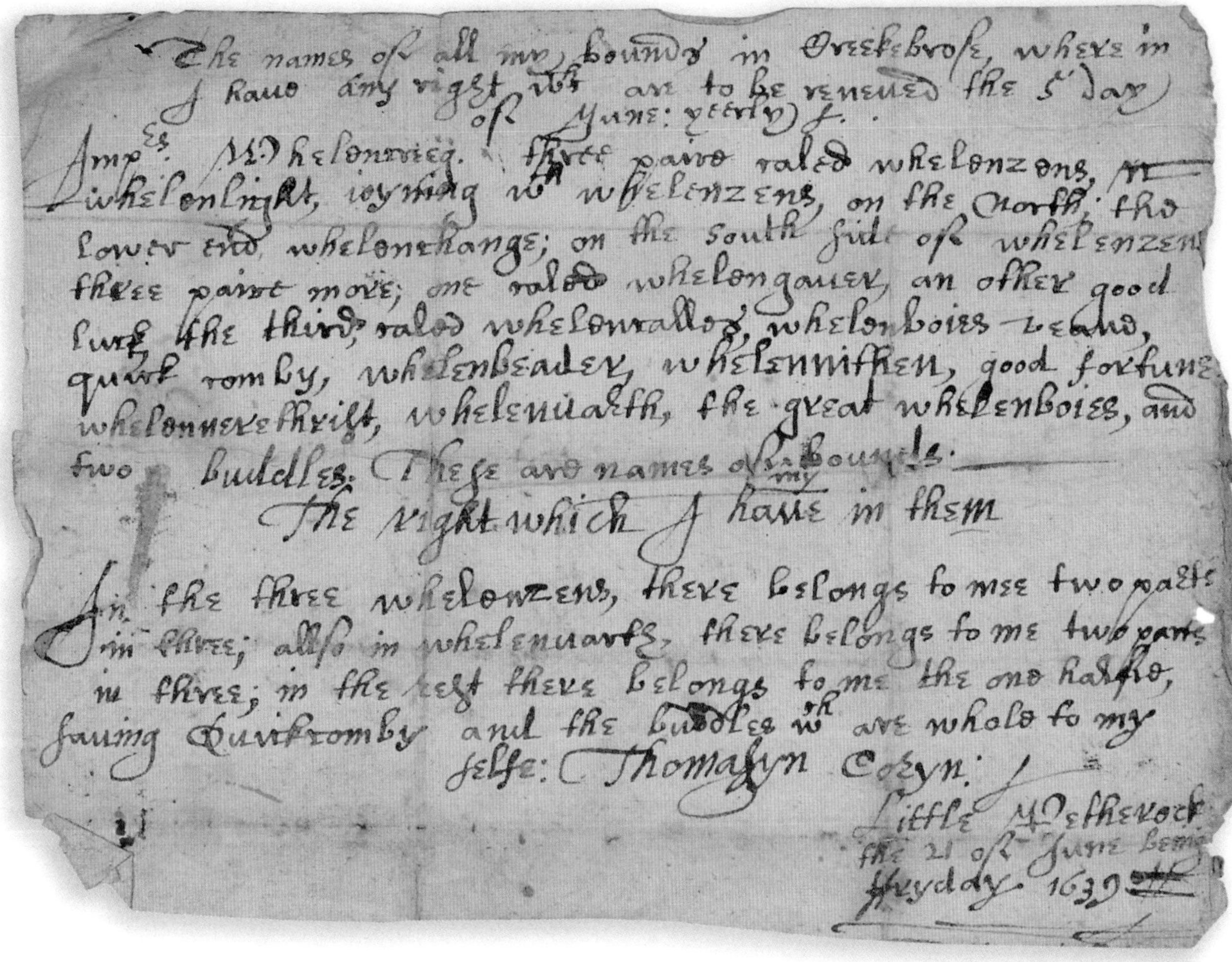
The names of all my bounds in Creekebrose, where in
I have any right wch are to be renewed the 5th day
of June: yeerly.
Impr. Whelenvieg. three paire caled whelenzens,
whelenlight, ioyning wth whelenzens, on the North: the
lower end, whelenchange; on the South side of whelenzens
three paire more; one caled whelengauer, an other good
luck the third, caled whelenvallos, whelenboies teaho,
quick romby, whelenbeader, whelennithen, good fortune
whelennere thrift, whelenuarth, the great whelenboies, and
two buddles. These are names of my bounds.
The right which I have in them
In the three whelenzens, there belongs to mee two parts
in three; allso in whelenuarth, there belongs to me two parts
in three; in the rest there belongs to me the one halfe,
saving Quickromby and the buddles wch are whole to my
selfe: Thomasyn Coryn:
Little Methevock
the 21 of June being
fryday. 1639

could expand on a hitherto unimagined scale. The East India Company was nearly 100 years old and the South Sea Company had yet to create the financial 'bubble' that was to burst so infamously. The National Debt was established in 1693, effectively giving the government a mechanism for borrowing from the wealthier classes – the merchants and landed gentry. Those making loans received interest at a guaranteed rate secured on the sound credit of Parliament. The Bank of England was established in 1694, helping to bring reliable financial arrangements to the expanding mining, smelting, manufacturing and trading industries of the country. The value of joint-stock capital in England doubled between 1695 and 1703, from £4,250,000 to £8,500,000. This represented 12 per cent of industrial wealth in England at that time. Of the 54 major joint-stock companies chartered between 1660 and 1719, 23 were concerned with mining, smelting and manufacturing.[13]

List of tin bounds owned, in 1639, by Thomasyn Coryn, widow of a wealthy gentleman. These tin bounds were on Creegbrawse, near Chacewater. The tin mines at Creegbrawse were worked from the medieval period right through to the early 20th century.

These improved arrangements for finance and company management were reflected at mine level here in Cornwall. Companies of adventurers who had joined together to obtain a mining lease and open and operate a mine were having more formal agreements and articles of association drawn up. The relatively informal arrangements by which tinners registered and worked their tin bounds under the old stannary system were largely being replaced – especially when a lode mine was opened – with legally binding articles drawn up by lawyers specialising in mining matters. Such agreements bound the shareholders to respect majority decisions, appoint and accept the financial management of pursers, share the costs and profits and forfeit their share if 'calls' were not paid within a specified time. (A 'call' of so much a share was made to adventurers when the mine needed financial support.) The agreement drawn up for Wheal Widden in Kea, dated 21 April 1684, was typical, containing as it did the names of the adventurers or shareholders, who were mostly local gentry or merchants, and detailing the way the company was to operate. Wheal Widden was worked intermittently from medieval times until 1991, when, as part of Wheal Jane mine, it finally closed.[14]

Another example of mines being operated with more formal agreements had to do with Carnkye Bal's new water engine, referred to earlier in this chapter. On 4 July 1699 three separate agreements were entered into by the mine owners. John Richards and seven partners were contracted to 'sinke down two engine shafts ... allotted and marked out ... [with] such dimensions in length (as) will hold a water engine.' The two shafts were each to contain five water-wheels of 12- to 15-ft (3.5- to 5-m) in diameter, and Richards and his partners were to equip the shaft with all necessary supports and appurtenances. They were to be paid £130 for the task. Next, Robert Buggens and partners were contracted to construct a water leat from Wheal Jowe adit, across Filtrick farm, to the Carnkye Bal engine pond. They were to be paid £210 for the job, plus £2 a year for its use. Finally, John and Thomas Tomson, engineers, were to be paid £500 for erecting an engine in the two shafts 'to draw water from the deep bottoms'. Parties to the agreements were the mineral lords, the adventurers and the bounders, together with the tinners who worked the adjacent mines and would benefit from the new drainage scheme. Until then the eight named tin bounds had worked as separate small, shallow mines, but thereafter they were to be gradually amalgamated into one efficient mine, eventually forming an important part of Basset Mines, which closed in 1917.[15]

Not only did the management of the mines change during the 17th century, but the attitude of the mineral owners also changed as they became aware of the value of having the minerals on their land made more easily accessible. When a mine was well run and the 'fixed plant' well maintained and left in good condition when its lease expired or the mine was abandoned, the value of the mineral lord's property was enhanced. Lease agreements began to reflect this new awareness and contained specific clauses detailing how shafts and other workings were to be left in good condition and with the timber supports sound and in place. The lease granted to a group of adventurers for 'Goneva worke alias Wall worke' in Gwinear well demonstrates this growing awareness. Dated 19 August 1699, it states that when the mine was abandoned on expiry of the lease the adventurers were not to 'riffle destroy remove take or carry away any of the ffixed tymber or stemples ... ffixed for support of the said worke or in any of the shafts'. Also, when engines or tackle were removed by the adventurers, the buildings that housed them were to be left in good condition. A crucial clause showed the mineral lord's understanding of the long-term value of the mine, for the adventurers were encouraged and given incentives to extend the adit drainage system. In the case of Goneva mine, the lords contracted to reduce the dues paid to them from a seventh to a ninth part once the adit was driven into the mine. Such a mine would make an attractive proposition to prospective adventurers.[16]

The century closes with a portentous discovery

As the 18th century approached, most of the old tin mines of Cornwall were facing a new problem. Below the water table their ore was increasingly found to contain in its 'poder' (waste) a material that was not recognised. Wolfram the mine operators did know about, for it had been a problem in some mines for many years, contaminating the ore and, because it has the same density and specific gravity as cassiterite (the principal tin ore), making separation difficult, but this other material was something the tin miners did not readily recognise. It was to prove a more valuable discovery than those seventeenth-century tin miners could possibly have imagined, for it was to bring wealth on a massive scale to Cornwall and change its landscape forever. The mysterious waste material was rich in copper.[17]

A view of the nineteenth-century coffin at Carclaze Tin Works, near St Austell, shows the extensive use of water power. This district produced tin from medieval times.

CHAPTER FIVE

COPPER: BIRTH TO BOOM

16th–18th centuries (1555–1775)

OPPOSITE
A view of the cliff workings at Portreath as they appear today. These old workings were part of Wheal Mary Mine and produced copper ore. The copper lode can be clearly seen as it approaches the surface above the worked-out stopes.

THERE IS NO CERTAINTY that copper was mined in Cornwall during the Bronze Age but, with the general demand for copper from at least the second millennium BC, it is hardly likely that the clearly visible copper lodes in the cliffs of west Cornwall would have been missed by tinners or merchants interested in raw materials for copper or bronze working. Even the Elizabethan cartographer John Norden, who was neither a geologist nor a metallurgist, noted that: 'The sea clyffes betwene this place [Pendeen in St Just] and St Ithes [St Ives] doe gliter as if ther were muche copper in them.' Not only is copper to be seen along that stretch of coastline but also between St Ives and Lelant, between Portreath and Perranporth and in several places in the cliffs of Mounts Bay. Such lodes would certainly have been noted by merchants approaching the shores of west Cornwall, as they would have been by local fishermen. Evidence of the making of bronze weapons in Cornwall, such as the Bodwen rapier mould from the Middle Bronze Age, supports this theory. As the richest of the outcropping tin lodes would have been exploited by the tinners, the skills and tools needed for hard-rock mining of copper ores would also have been available. The Romans mined copper in Wales, though they have left no record of having done so in Cornwall. Despite the lack of written evidence, it is inconceivable that no exploitation of this easily obtainable copper ore took place between the Bronze Age and the Saxon period in Cornwall.[1]

The crown and the search for metals in the 15th and 16th centuries

Throughout the late medieval period the English crown granted licences to search for various metals – including copper, silver and gold – in several counties. From the time of Edward I onward these licences were granted to encourage the search for metals of use to the country and by which the crown could obtain valuable duty. In 1452 Henry VI made such a grant to search for various metals, including copper, in Cornwall: 'The King by his Letters Patents makes his Chaplain John Bottwright Comptroller of all his Mines of Gold and Silver, Copper, Latten (brass), and Copper Latten Lead, within these two Counties [Devon and Cornwall].' Copper was found and exploited in other counties, and although little worth mining was discovered in

Cornwall, mining men were aware of its existence and potential value. By the end of Henry VIII's reign (1509–47), industrialists were once more turning their attention to the lodes of Cornwall as potential sources of copper supply.[2]

In 1555, during the reign of Mary(1553–58), the government gave permission for a German mining engineer, Burchard Cranych, to search for copper in Cornwall. What he found was hopeful, but the enterprise was not a success. A few years later, in the early part of Elizabeth I's reign (1558–1603), a company called the Mines Royal was set up specifically to search for, mine and smelt any metallic ores as yet undiscovered and exploited. The company was incorporated by the crown in 1564 with a board of half English and half German directors and managers. The German Daniel Hochstetter came to Cornwall to direct operations, but once again they were not successful. Twenty years later a more determined attempt was made, and this time there were definite areas of success. Ulrich Frosse, another German mining engineer, led the work and copper was discovered in the cliffs at St Ives, St Just, Illogan, St Agnes and Zennor. Other copper lodes were discovered at Marazion and in an old tin mine at St Hilary, just along the coast. However, despite the fact that copper was his main interest, Frosse spent most of his time mining for lead and silver at Treworthie mine, between St Cubert and Perranzabuloe. Lead was also mined at Porthleven. Many of these mines had been originally worked by Hochstetter and Cranych, and Frosse reopened them. Although some German miners worked in Cornwall, most were sent by the Mines Royal to Keswick to open up the copper mines there, where the whole operation from mining and dressing to smelting the ore worked well. The Cornish ore was sent to Neath, in South Wales, where the Mines Royal had a smelter, although some ore is believed to have been part-smelted in Cornwall.[3]

None of the three German-led attempts to exploit the copper lodes of Cornwall was successful. Although the work done in the 1580s located good copper ore and produced enough to transport to Neath, in general they produced insufficient good-grade ore for the operation to pay for itself and make a profit. The much-vaunted German technology was not efficient. Their rag-and-chain pumps could not cope with the water even at the shallow levels at which they were working, and although they drove several adits for drainage, the rubbish left from the previous workings caused constant problems, not least because water seeped from flooded workings above the new adits. Frosse lacked the required energy for such a task, and he constantly complained about everything from his health (he was a hypochondriac) to his fear of heights (he had to climb down ladders on cliffs) and the lack of sufficient funds to pay the miners and buy gear. To make matters worse, the Cornish miners taunted Frosse with the claim that they were better workers and more skilled miners than him and his Germans. Within a few years the Mines Royal had pulled out of

Cornwall and nothing more was done to exploit the copper lodes in the Duchy for nearly a century.[4]

The letters between Ulrich Frosse and the various shareholders in the Mines Royal tell us a considerable amount about how these mines were worked. Treworthie mine, which was worked for lead and silver, has left a detailed description of its organisation, costs and working methods. A typical crew consisted of ten men: three pickmen, who broke the ground at the heading; two barrow men who moved the broken rock back to the shaft; two men winding the windlass which hauled the rock up to surface; two pumpmen operating the rag-and-chain pump; and one mine captain, who also took charge of the support timbering. Another letter speaks of 13 production miners split into two 12-hour shifts. The adits, most of which were being driven beneath old, flooded workings, were about 7 ft (2 m) high by 2 ft (1/2 m) wide. They had to be timbered for their entire length as the ground was soft and given to collapse. The miners were paid 3s 6d a week and the mine captain 5s. Tools,

This evocative, but perhaps idealised, drawing of 1814 depicts mining at St Agnes in the early 19th century.

gear and the clerk's wages came to £1 a week, which totalled £2 16s 6d. If the water was particularly heavy, Frosse reckoned that the weekly bill would rise to about £5 10s. When this happened, three men would wind the pump for 12 hours and another three for the next 12 hours. The miners took turns at the various tasks rather than staying continuously on one job. Due to the soft ground, impressive advances of up to 2 ft (1/2 m) a shift were sometimes attained, although the timbering work would have inhibited progress to some extent. Though operated on a smaller scale, the dozen or so other mines probably worked to the same system. It is remarkable how similar the working practice at Treworthie was to the description 80 years later of the tin mines on Hingston Down.[5]

This beautiful specimen is from the collection of Philip Rashleigh at Truro Museum. It is described as a 'natural solution of copper attracted by iron ore and forming radiated malachites with iron ore'. It came from Wheal Husband Mine in Gwinear in the late 18th century.

1680s – the mines strike copper

It was in the 1680s that copper again came to the attention of Cornish miners. As the tin mines deepened below the water table, more and more copper ore was being brought to surface in the waste. The timing of this was perfect, for the country was changing in a number of ways, all of which made the discovery of so much copper ore extremely significant for Cornwall. It was not just the constitution that altered with the Revolution of 1688–89; the whole of society changed, as did the economy, politics and industry. The late 17th century saw intense activity as old industries expanded, new industries were created, new inventions and innovations were introduced and joint-stock capital became available to finance enterprise as never before. The National Debt and the establishment of the Bank of England changed the attitude of country landowners to finance and investment, and soon their surplus money was capitalising mines, factories and smelters. As we have seen, improved mining techniques together with better pumping and smelting technology were making mining and smelting far more efficient, and mine owners could view the future with confidence. Among the first reforms that the revolutionary government made was the abolition of the Mines Royal monopoly, freeing mine owners and mineral lords to exploit the minerals on their land. Coinciding with these changes was the demise of Sweden as England's main supplier of copper. Copper was a central raw material in the successful expansion of industry, not just for itself, but as the main constituent of brass and bronze. The zinc ore calamine, used in brass-making, had been discovered in the Mendips in the 16th century, and by the end of the 17th it was being mined again for use in the burgeoning brass foundries of Bristol.[6]

William Pryce described the rediscovery of Cornish copper: 'About eighty years since [i.e. in the 1690s], some gentlemen of Bristol made it their business to inspect our mines more narrowly, and bought the Copper Ores for two pounds ten shillings to four pounds per ton. The gains were … so great, that they could not long be kept secret, which encouraged the other gentlemen of Bristol about sixty years since to covenant with some principal Miners in Cornwall to buy all their Copper Ores for a term of years at a stated low price.' Mr Beauchamp of Trevince sold ore from his mines for £5 a ton for a period of 20 years. The adventurers of Relistian sold theirs for only £2 10s a ton. Although they were clearly taking advantage of the Cornish, these copper agents and the Bristol-based companies they represented were largely responsible for the establishment of Cornwall as the principal source of world copper for the next 150 years. Men like John Coster, Sir Clement Clarke, Sir Ambrose Elton and Grabriel Wayne brought not only necessary capital into the copper mining industry, but also business know-how and the contacts that made the industry a success. Coster was to be the dominant figure in Cornish copper mining for decades, and his sons carried on where he left off.[7]

Early 1700s – vertical integration

By 1700 an increasing tonnage of copper ore was leaving Cornish ports for Bristol, where the new reverberatory furnaces were using coal to smelt the copper. With coal and calamine (a zinc ore) close to hand, and capitalists such as Clarke and Elton to back the projects, the success of the Cornish copper industry seemed assured. Brass foundries were set up close to the smelters, and links were established with manufacturers in the Midlands, ensuring them a steady supply of copper and brass. At the same time Coster, Wayne and other agents took their plans a stage further by acquiring shares in the mines themselves, sometimes even becoming principal adventurers. Their names began to appear on the leases granted to mining companies, especially where copper ore was found. These Bristol-based copper agents became important shareholders in Goveva mine, Gwinear, Pool mine, Illogan, Chacewater mine, Kenwyn, Wheal Owles, Camborne, and North Downs, Redruth. Early on there were just four copper companies that bought Cornish copper: the Brass Wire Company, Wayne and Company, Chambers and Company and the English Copper Company. The English Copper Company was the largest and tended to dominate the market. By the end of the 17th century Coster's company and the English Copper Company, established in 1691, were sending nearly 1000 tons of copper to their Gloucestershire smelters every year. Within a couple of decades there was hardly a copper-producing mine in Cornwall where these outsiders were not involved. They gradually gained control of the copper industry, producing the ore, transporting it to Bristol, smelting it and supplying the finished product to the brass foundries and the manufacturers. Their anxiety to control Cornish copper was also motivated by the resurgence of copper smelting in South Wales. The old Mines Royal had maintained its own smelter at Neath, and in 1717 this was joined by furnaces established at Swansea. The Welsh had the advantage of being a day's sailing closer to ports on the north coast of Cornwall than Bristol and Chepstow and, despite the best efforts of the Bristol-based copper companies, this proximity was eventually to prove decisive. The year 1717 also saw, for the first time, use by the English mint of British copper for its coinage, most of it originating in Cornwall.[8]

The ticketing system

When the first Welsh copper agent visited Cornwall, several hundred tons of copper ore were lying unsold at Roskear and Wheal Kitty mines in Camborne. The four 'confederated' companies would offer no more than £4 5s a ton, and the adventurers were reluctant to sell it. The Welsh took 1400 tons at £6 5s a ton and another 900 tons at £7 a ton. According to Pryce, these smelters still made more than 30 per cent profit even at those fairly elevated prices. By price-fixing, the Bristol smelters had kept the price low, but with this new competition the mines were able to operate more profitably. The problem of collusion among the copper agents was not solved by competition from Wales as it was not long before they joined in the arrangement, so a method was devised to ensure that all ore was sold and the mines were given a fair price. The long-term answer to the problem – known as 'ticketing' – was typically Cornish. Ticketing began in 1725 and involved piles of ore ('parcels') being sampled by smelters' agents, who made bids for each parcel of ore and placed them in sealed envelopes. The ore went to the highest bidder. All parcels of ore were sold, with the responsibility for assaying on the buyer. Once collusion again became a problem, neutral 'referee' assayers were introduced. The average grade of dressed copper ore sent to the smelters was about 7 per cent. Some ore continued to be sold by private treaty, but many lease agreements stated that copper ore must be sold by ticketing as mineral lords were keen that all ore was sold and their dues paid promptly. As to tin, with the ending of coinage in 1838 it could be smelted anywhere, not just within the appropriate stannary district as formerly. Importantly, it also meant that tin ore was available for ticketing, and for a large part of the 19th century most Cornish tin was sold by this system.[9]

Another response to external control – smelting in Cornwall

No sooner had the agents of the Bristol smelting houses discovered the value of Cornish copper in the 1680s than attempts were made to smelt the ore in Cornwall. In 1693, before he became involved in copper smelting in Gloucestershire, Sir Clement Clarke and his son Sir Talbot had associated with Francis Scobell, MP for Mitchell, Henry Vincent, MP for Truro, and the Carlyons of Tregrehan to set up a copper smelter at Poldudden, near St Austell. This London-based company lasted only four years before closing due to poor management, dishonesty and lack of easily available ore to smelt. Its distance from the principal copper mines was the main reason for its failure, and attempts to buy copper as far away as Chacewater were unsuccessful. Sir Clement moved to Bristol and established furnaces there, employing Gabriel Wayne and John Coster as agents. In Bristol he could take advantage of the lower transport costs of coal from local sources and from South Wales, as smelting required this in some quantity. However, Cornishmen continued their efforts to smelt copper in the county, and in 1696 John Pollard of Redruth, together with Thomas Worth of Penryn, set up a smelter at St Ives. It was probably their smelter that Celia Fiennes saw when she visited St Ives in 1698. It was not a success for technical reasons, but in 1712 they tried again with relative success for ten years. Pollard then moved to Swansea to take on the Welsh on their home ground. Meanwhile, in 1710, sea-borne coal duty was abolished, which encouraged Cornish attempts to smelt copper in Cornwall. Another smelter was set up by Gideon Cozier at Phillack in 1710, close to the Hayle coal quays, and when he died Sir William Pendarves, a major mine owner, took over, later to be joined by John Coster's son Robert. The firm worked successfully for 25 years and only closed when the principal shareholders died. It was financially viable due its ability to obtain ore from mines that were not controlled by the up-country copper companies, several of them on Pendarves land. Gideon Cozier's brother, Thomas, apparently operated a tin smelter at Perranarworthal between 1721 and 1724; this had four reverberatory furnaces, which he is thought to have used also to smelt copper ores.

Other experiments in copper smelting were carried out at Trewinnard House, St Erth, in the late 17th century. Undoubtedly, the most successful copper-smelting company in Cornwall was the Cornish Copper Company. Sampson Swaine established furnaces on Rosewarne Downs in Camborne in the 1740s, and in 1748 he was invited by the adventurers of Dolcoath and Bullen Garden to set up furnaces at Entral, beside Bullen Garden mine. The smelting was a success, but the distance from the coal ports was considerable, and in 1756 the whole operation moved to what became the Cornish Copper Company at Copperhouse, Hayle. The company was officially formed in 1758 with capital of £22,000, mostly raised by Dolcoath adventurers from Camborne. Despite continued and often bitter attacks on the

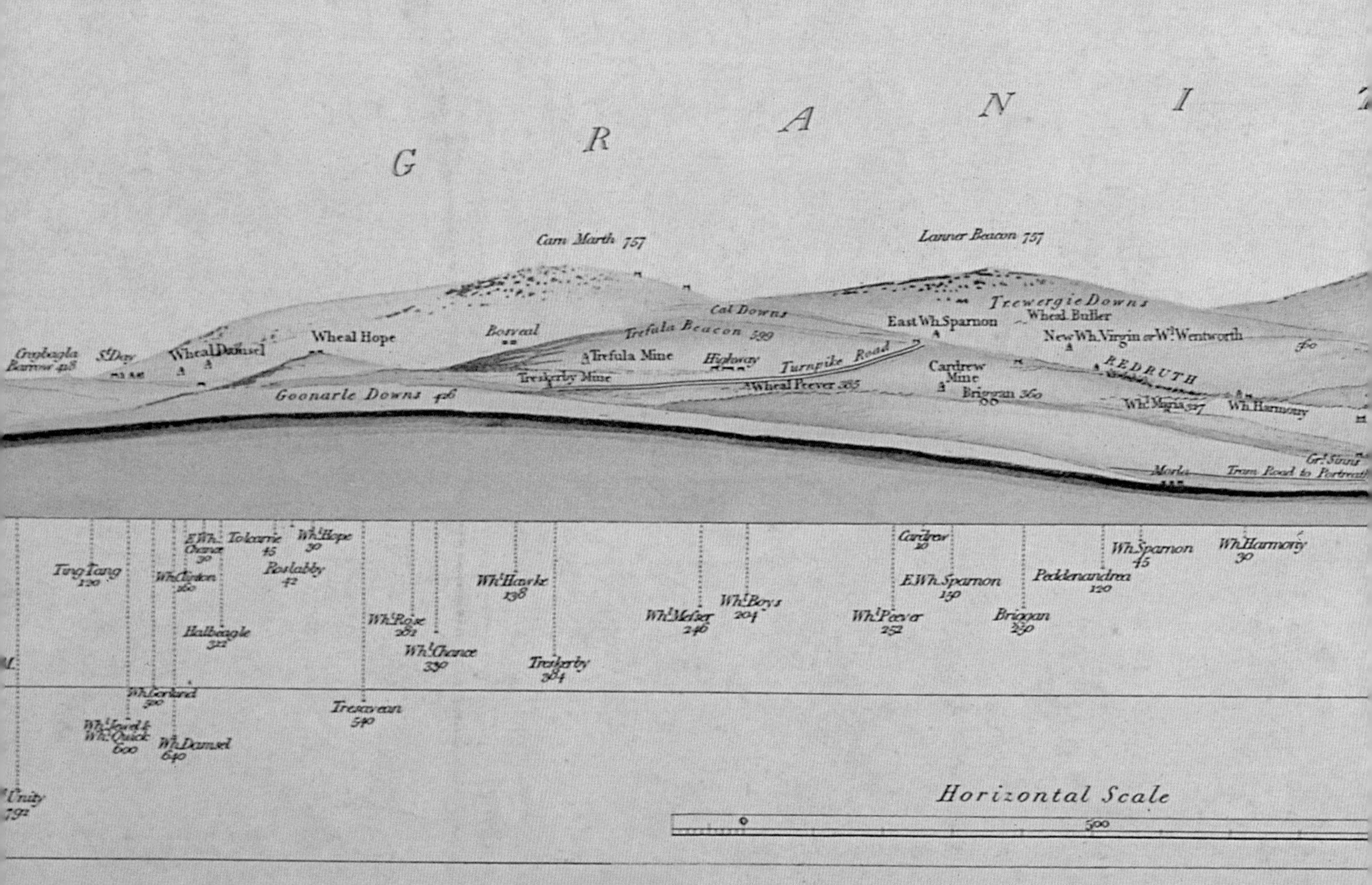

An 1819 drawing by Richard Thomas of the mines of the Chacewater to Camborne mining district. The section shows the height of mines above sea level, and the depths of the mines.

…as which overlays it has been found to dip regularly …In some of the Mines, Granite and Killas are found …r places detached irregular masses of Granite are found …ear the main body of the Granite.

The whole of this ground is traversed by Copper and Tin Lodes and Elvan or Porphyry dyk… between 20 and 40 degrees S. of West and N. of East; and by Cross Courses composed princ… tions of which are generally between 15 and 35 degrees W. of North & E. of South; all of which… depth and continue their courses uninterrupted by the various Strata _ Elvan is also foun… irregular masses _ Here and there are masses of Iron Stone which though apparently irregular … a general course nearly as the Lodes.

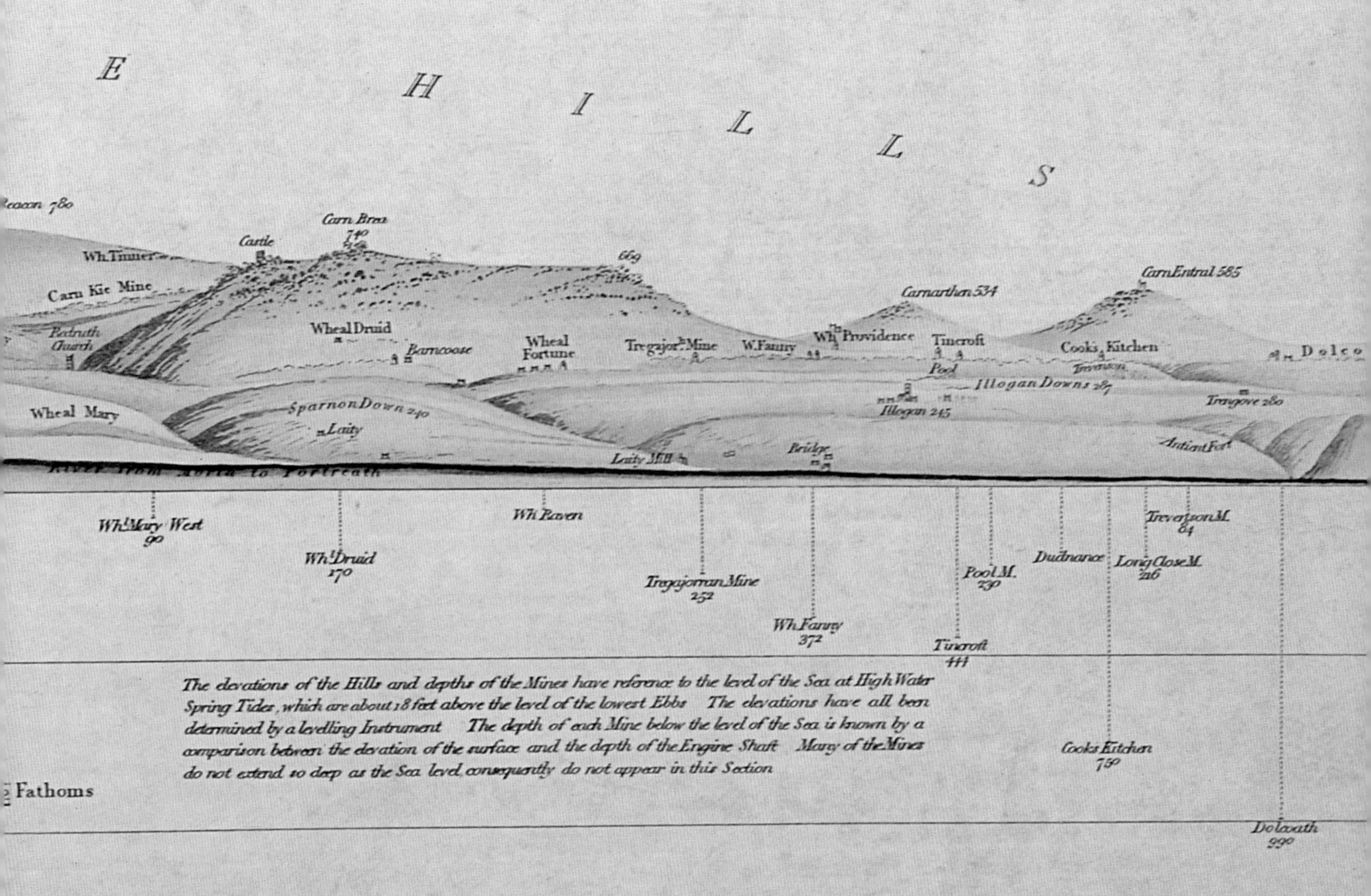

…G WITH THE MAP FROM CHASEWATER TO CAMBORNE,

by R. Thomas. *Falmouth 1819.*

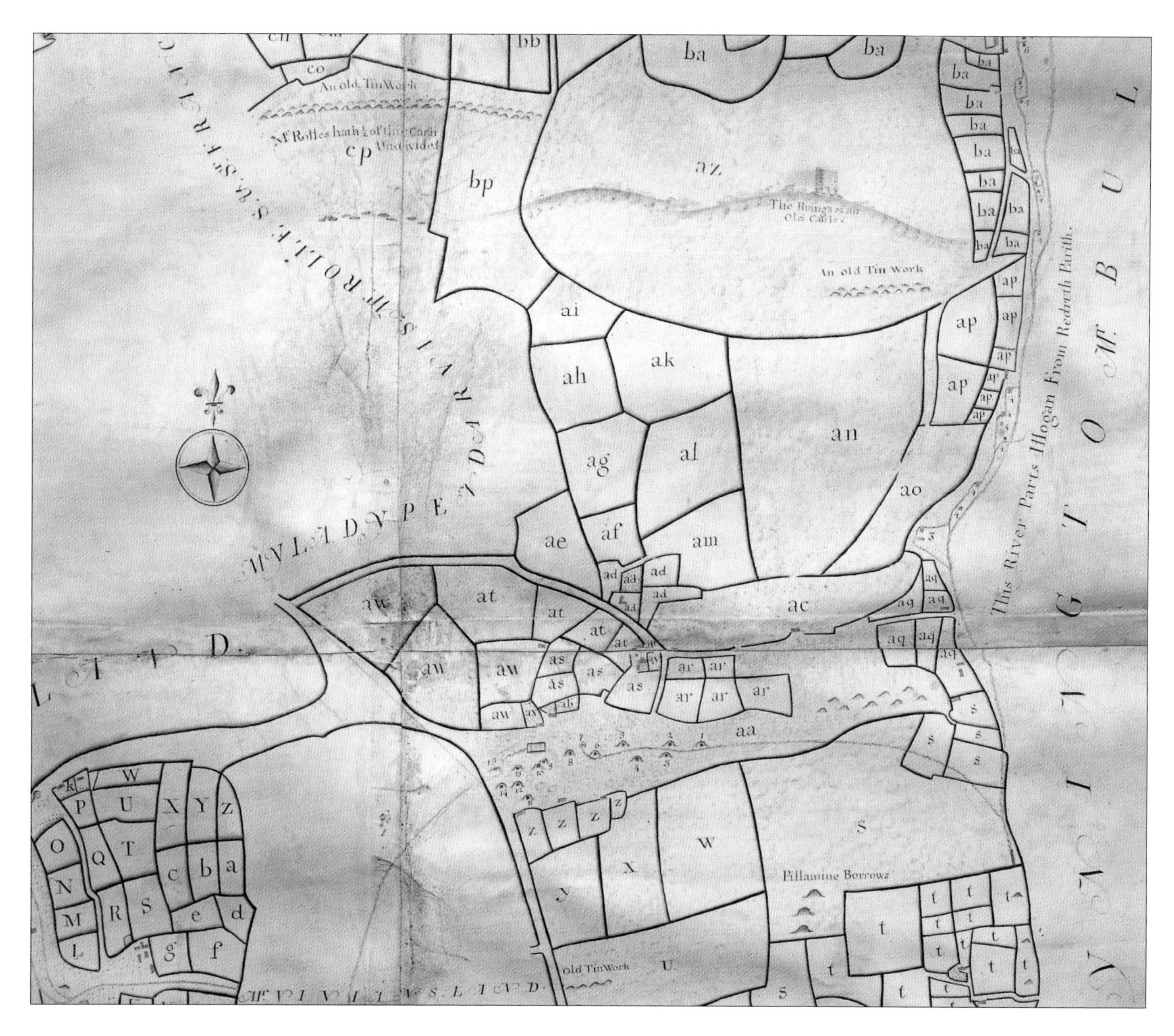

A detail from one of William Doidge's maps of Tehidy Manor, drawn in 1737. Across the centre are the ancient shafts of Carnkye Bal, which from the early 16th century was one of the most important tin mine groups in Cornwall. In the late 19th and early 20th centuries this old bal formed the centre of the enormous Basset Mines Group.

company by the Welsh smelting companies, it successfully continued to smelt Cornish copper ore for more than 60 years, until about 1819.[10]

Henric Kalmeter, a Swedish industrial spy, visited Cornwall in 1724 and described the many copper mines he visited. He commented that in general copper mines had originally been tin mines, and that the red metal was found beneath the white. This was certainly true of the mines in the Camborne, Redruth, Gwennap and Chacewater areas, where many ancient tin mines had become prosperous copper mines. In some cases Coster and the Bristol company reopened abandoned tin mines themselves in their search for new sources of copper. An example of this was Wheal Lovely, 'first worked for tin and then abandoned', which was taken up by Coster and company in 1716 to work 'principally for copper'. Writing in the 1750s, William Borlase noted that between about 1715 and 1755 some 18 copper mines had dominated production in Cornwall, fourteen of which, in the Camborne–Redruth area, were former tin mines. Pryce wrote that in the 1720s three of these mines, Wheal Fortune in Ludgvan, Roskear in Camborne and Pool Adit in Illogan, produced 'great quantities of Copper Ore'. The increase in copper production was spectacular. From something less than 1000 tons in 1700, by the mid-1720s an annual average of 6480 tons of copper ore, worth some £47,350 to the mines, was being shipped to the smelters.[11]

The tribute and tutwork systems

During the first part of the 18th century the tribute and tutwork systems were developed, and they have become closely associated with Cornish mining. There has long been confusion about how these systems operated in Cornish mines, and the single-sentence glossary definitions found in a variety of books have not helped. Both practices evolved through the developing arrangements necessary for the new, highly capitalised and rapidly expanding copper mining industry. They grew from small tin mines where adventurers often worked the mines themselves. It was a small step from being a working adventurer who shared the value (and the costs) of the ore raised to being an independent tributer who did not receive a wage, but rather a share of the value of the ore and paid his own costs. Many of the wage-earners became tutworkers (piece workers), or worked on the 'owners' account', being paid a daily or monthly wage. Tribute pitches (designated parts of a stope) were set by means of a Dutch auction, where miners offered to work a pitch at a 'tribute' of so much in the pound. The lowest bid usually secured the pitch, provided that the mine captain or manager thought the mine would get a fair return and the tributer could do the job. Pitches were set for periods of one to four months, most lasting for two or three months. If the ore grade in the stope was particularly good, the tribute might be as little as a shilling in the pound, but if the ore was not much good and the prospects poor, the tribute paid to the miner could be as high as 19s 6d. A normal high tribute would be up to 13s 4d, and a low one not usually less than 5s in the pound.[12]

It has been usual to regard the tributer as more skilled than a tutworker, but that was far from the case. A tributer's skill lay in his ability to identify good-grade ore and break it with as little dilution as possible. He would organise and supervise a group or 'pare' of miners, which could number anything from six to sixty, and these worked away the ground in their pitch. The actual mining skill involved in stoping away the ground was minimal, for the tributer's skill lay in knowing what to remove. On the other hand, the tutworker's skill lay in his ability to break ground quickly and efficiently. As almost all tunnels (levels) were driven by tutworkers and all shafts and winzes were sunk by them, they had to have considerable skill in ground-breaking techniques that were not required of the tributer. To advance an end or sink a shaft, the miner had to blast to a 'free face', as such a face is needed when explosives are used to break rock. This involves drilling the holes that are to receive the gunpowder in a pattern whereby the rock will be blasted away from the rock face. Every rock face needs a slightly different pattern of holes. Thus, the tutworker's skill was not inferior to that of the tributer but different.[13]

Another widely believed myth is that tributers and tutworkers were different men. Examination of scores of mine cost books show that as miners moved from mine to mine they also moved from tutwork contracts to tribute contracts and back again. Some miners stayed at Dolcoath for years as tributers, then moved next door to Cooks Kitchen as tutworkers, staying a few months before returning to Dolcoath as tutworkers, and later moving elsewhere to work on the 'owners' account' for a daily wage. A tributer, or 'taker' (the miner who 'took' the contract), might divide his monthly 'gettings' between his men equally, but more usually he would pay his men an agreed amount, perhaps £1 10s for the month, and take the rest for himself after paying his costs to the mine. There might be two or three 'takers', who shared the tribute contract and supervised the work. The mine would deduct from the tributer's takings the cost of tools, smith's charges, candles, gunpowder and the amount charged by the mine for drawing the ore to surface, dividing and assaying it. The tributer might receive a large sum if the ore was good, but he might also end up with little or nothing after his men were paid and costs were taken out.[14]

The tutworker would have similar charges, although he would not have to pay for the dividing and assaying of ore. Due to the nature of the work, the tutwork contract involved small groups of men. Two men might drill and blast the end and another, usually a boy, would barrow the broken ground back to the shaft for hoisting to surface. Tutwork was paid at so much a fathom, and the price reflected the difficulty involved. If the end they were driving was very hot or the distance to the shaft considerable, the price could be high. If it was a cool end and the shaft nearby, the price might be low. The hardness of the ground or the need for timber supports if it was weak could also affect the price paid per fathom. The same mine might pay as much as £20 a fathom to one tutworker in a deep, hot and remote part of the mine, and as little as 4s to another who was driving a cool, shallow level near the shaft. Most tutwork contracts gave a price per fathom up to so many fathoms, and, if the miners advanced to that point before the end of the month's contract, the mine captain might reduce the price until the end of the month. Exceptions to this were when the mine needed a tunnel to be driven as fast as possible, as when bringing an adit home to the mine. Another group of miners was often referred to in cost books as 'stopers'. They were, in effect, tutworkers doing the job that was usually done by tributers. They worked in stopes, breaking the ground as did tributers, but they were pieceworkers paid at so much per cubic fathom of ground broken. The stopes they worked were called 'owners'' stopes and the costs were borne by the adventurers.

When tutworkers were driving an end 'on lode' they might be paid a proportion of the value of the ore as well as for the fathoms they had advanced. What must be emphasised is that during the 18th and 19th centuries no two mines worked in precisely the same way, and no mine employed the same system uniformly for any length of time. Every tutwork and tribute contract in every mine in Cornwall could be modified to suit local conditions, the prejudice of the mine captain or the skill and number of miners available. A former mine captain at South Crofty mine, where the contract systems had evolved over 300 years, remarked that the only word to describe the variety and perverseness of the multitude of formal and informal arrangements there was 'Byzantine'. If that was true of a modern mine in the 1990s, how much more was it true of mines in the first half of the 18th century.[15]

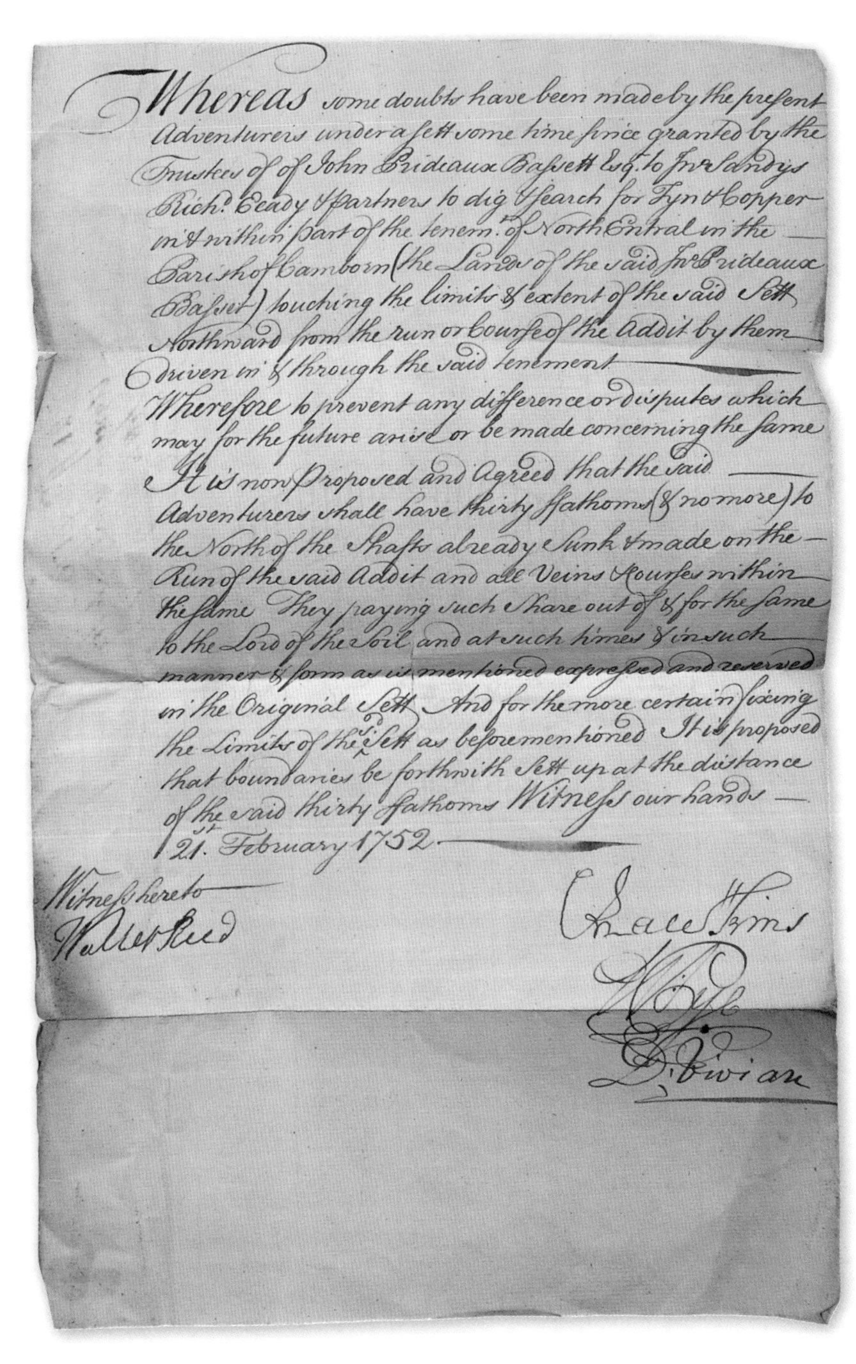

Whereas some doubts have been made by the present Adventurers under a sett some time since granted by the Trustees of of John Prideaux Bassett Esq. to Mr Sandys Richd. Ceady & Partners to dig & search for Tyn & Copper in & within part of the tenemt. of North Entral in the Parish of Camborn (the Lands of the said Mr Prideaux Basset) touching the limits & extent of the said Sett Northward from the run or Course of the Addit by them driven in & through the said tenement — Wherefore to prevent any difference or disputes which may for the future arise or be made concerning the same It is now Proposed and Agreed that the said Adventurers shall have thirty fathoms (& no more) to the North of the Shafts already Sunk & made on the Run of the said Addit and all Veins & Courses within the same They paying such Share out of & for the same to the Lord of the Soil and at such times & in such manner & form as is mentioned expressed and reserved in the Original Sett And for the more certain fixing the Limits of the sd Sett as beforementioned It is proposed that boundaries be forthwith Sett up at the distance of the said thirty fathoms Witness our hands —
21st February 1752

Witness hereto
Walter Reed

[illegible]
[illegible]
[illegible] Vivian

This 1752 agreement was drawn up to settle a dispute between two mines at North Entrall (next to old Dolcoath). It limited miners to working within 30 fathoms of the adit that drained their workings. North Entrall lay on the west side of the Red River in Camborne.

Another type of tributing that contributed to the success of Cornish copper mining was when a mine was 'sett on tribute'. Pryce described it in these words: 'The manner of setting or leasing a Mine on Tribute, is this; some able Miner takes the Mine of the adventurers for a determined time, that is, half a year, a whole year, nay even seven years, as was the case of Bullen-Garden, and the means of her discovery.' The man who took Bullen-Garden on tribute was Henry John, and the way in which he worked was described by Henric Kalmeter, the Swede, who visited Camborne in 1724. Kalmeter visited Condurrow mine, where Sir William Pendarves was landowner and mineral lord and held nearly half of the shares in the mine. Rather than work the mine themselves, the adventurers leased it out to Henry John on tribute. John was to pay the adventurers 5s for every ton of copper ore raised, together with Sir William's share of the costs of the mine. When John took Bullen Garden mine on tribute for seven years, he discovered and opened up the greatest copper and tin ore lode ever

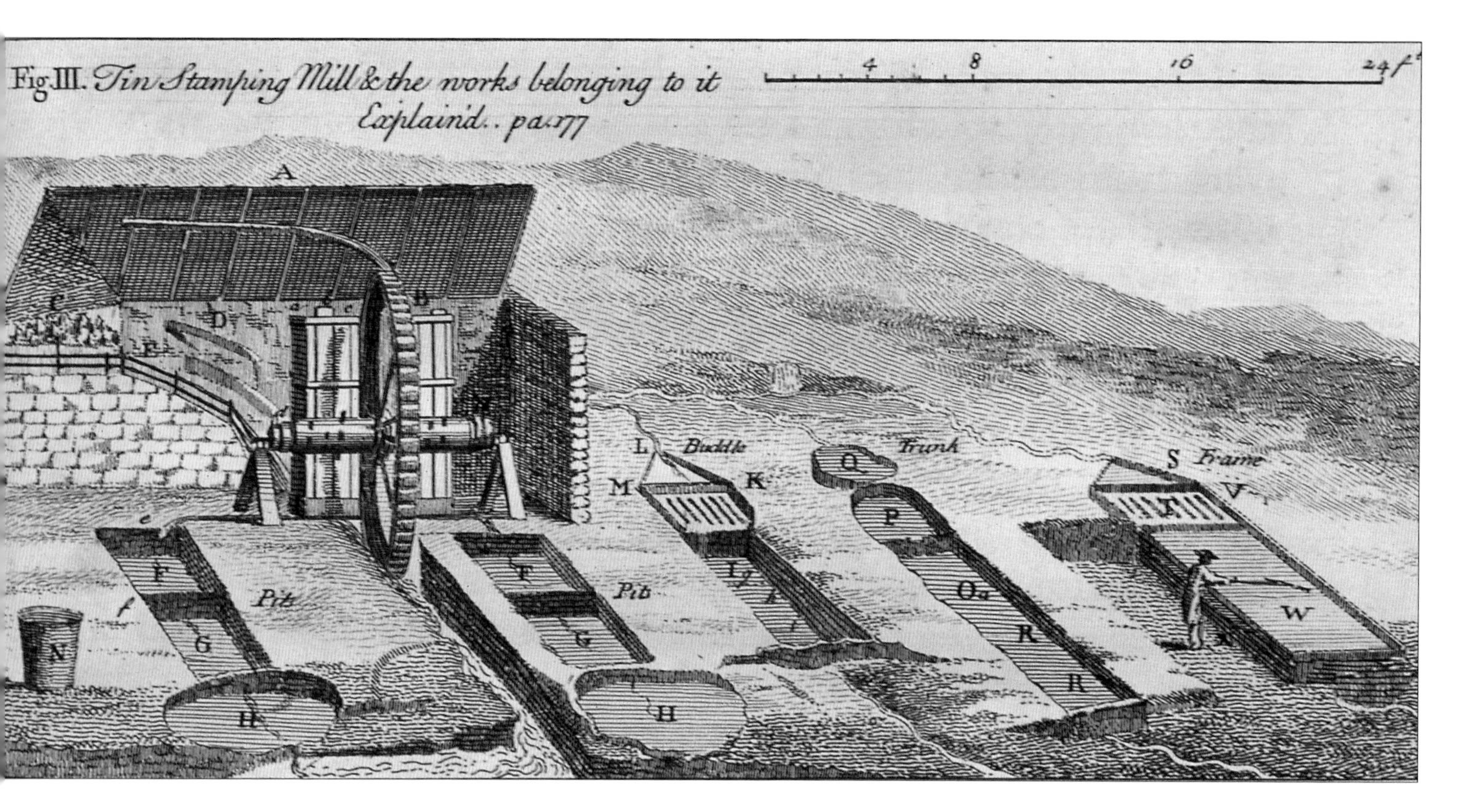

Tin stamping mill and concentrating plant as illustrated by William Borlase in 1758. The stamps are arranged on either side of the overshot water-wheel. The stamped material was then run into various pieces of equipment called a pit, a buddle, a trunk and a frame – which all used water to wash away the unwanted gangue from the cassiterite.

worked in Cornwall. This method of setting a mine on tribute continued into the second half of the 19th century. In 1862 South Wheal Crofty took the North Tincroft Sett from Tincroft mine and paid it a tribute of so much per ton of ore raised in the sett. It remained part of Crofty till the mine closed in 1998.[16]

Developments in response to larger, deeper mines

The system of working which had been developed in the 17th century whereby shafts were sunk, levels driven between them and the ground between the levels stoped away continued throughout the 18th century. With the greater depths and the need to remove much larger tonnages relatively quickly, mine managers had to develop the most efficient and economical ways of working. Exploration and deepening took place alongside production stoping, so that tunnels were driven on lode, crosscuts were driven between lodes, shafts and winzes were sunk for access and ventilation, and pitches were mined by tributers in a continuous and well-organised operation. Cost books and mine plans from the period show how sophisticated mine management had become. The plan and section of Polgooth mine, drawn in the 1690s, those of Pool mine, drawn in 1746, and the long section of Bullen Garden mine, drawn in the 1760s, all confirm this. Examination of these drawings show that mines were being worked to a well thought-out plan. Shafts for pumping, access and hoisting, adits for drainage, levels for access and haulage of ore, and stopes where the ore was being mined were all laid out to a practical and efficient design. The Pool Adit cost book for the years 1712–30 shows that tutworkers were driving through very hard rock at considerable speed. Henry Mill and John Paull and their pare drove an adit an average of 4 ft (1.2 m) a day in July and August 1712. At the same time miners were stoping the backs of the levels ('back stoping') and mining below the levels ('underhand stoping'). Water was brought to the mine by a leat more than a mile long to power the water engine, which was pumping water up to adit level. The mine employed a hydraulic engineer, George John, to maintain its water engine. This ancient tin mine was being rapidly opened up for copper, and John Coster and the Bristol smelting company were very much involved there. Scores of cost books and mine plans from the period show that mining techniques attributed by nineteenth-century writers to the late 18th century, and which they thought had been introduced from abroad, were in use by the start of the century.[17]

When William Pryce described the primitive, multi-wheel water engines of the late 17th century, he went on to mention the vast improvement brought to mine pumping by the introduction of an invention first patented in 1714. 'Mr John Costar, of Bristol, came into this country, and taught the natives an improvement in this machinery, by demolishing these petit engines, and substituting one large wheel of between thirty and forty feet diameter in their stead.' These engines were gradually

In 1862 South Wheal Crofty took the sett off Tincroft Mine adventurers on a tribute agreement, in which Crofty paid Tincroft a set amount for every ton of ore raised. This was a late example of an eighteenth-century system of 'letting a mine on tribute'.

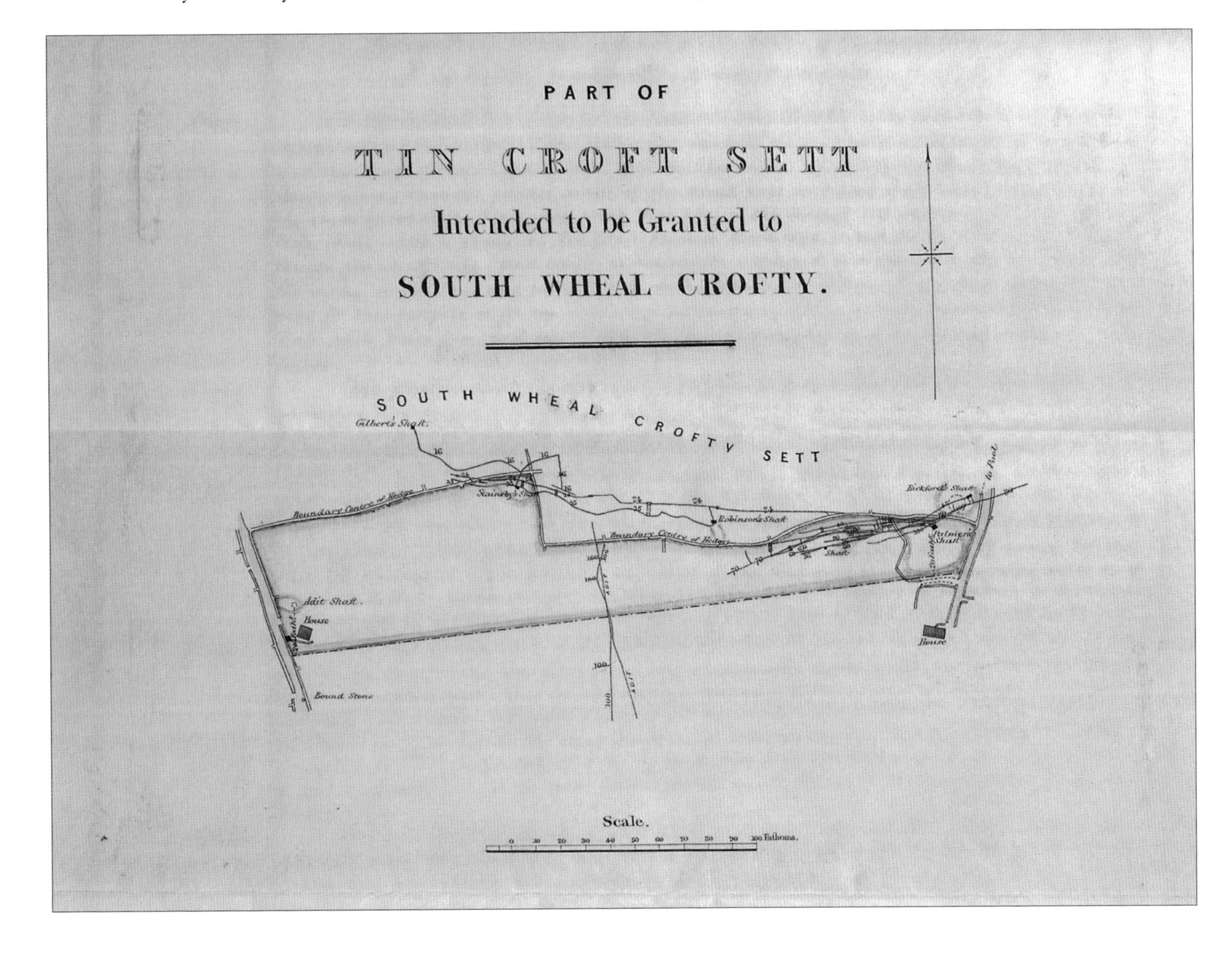

improved until they could pump efficiently from depths of well over 500 ft (151 m); indeed, Pryce cited the pump at Cooks Kitchen with a 48-ft (14.5-m) diameter wheel and pumped from 480 ft (145 m) below adit level. He believed with a better water supply it could easily lift water from 720 ft (218 m) below the adit. Coster's improved engines also operated with two wheels, one pumping with a suction lift, the other driving down a plunger. One of these engines was erected at Carnkye Bal in 1743 to replace the one installed in 1699. Walter Reed was contracted 'To erect and build with ye utmost Expedition two engines to draw ye Water as deep as ye ten Wheels did formerly. The Wheels to be thirty-three Feet Diameter.'[18]

Steam-powered pumping

Although for most Cornish mines water power remained the principal source of energy, the introduction of steam power was to revolutionise mining in the larger and deeper mines. As early as 1678 engineers had envisaged atmospheric steam engines, and by 1698 Thomas Savery, a military engineer, had patented a steam-activated pumping engine. It did not work well, but it laid the foundation for the work of Thomas Newcomen, whose atmospheric engine did work. In 1712 Newcomen erected an engine on a colliery in the Midlands, and it was not long before one of his steam engines was working on a Cornish mine. It is believed that the first Newcomen engine to be used in Cornwall was erected in 1716 at either Wheal Vor

Mine ventilation was always a problem. In 1778 William Pryce illustrated one successful method of moving the air through to a tunnel end, by creating a circuit using a false floor or 'sollar'.

HOW NEWCOMEN'S ENGINE WORKED

Newcomen's atmospheric engine consisted of a cylinder with a piston, which was connected to one end of a large overhead lever (beam/bob). The other end was fitted to a bucket pump in a shaft. A vacuum was created in the cylinder by filling it with steam and condensing the steam by means of a water jet. Atmospheric pressure then forced the piston downwards, pulling down the end of the lever over the cylinder and causing the end of the lever over the shaft to lift, which pulled the pump in the shaft upwards.

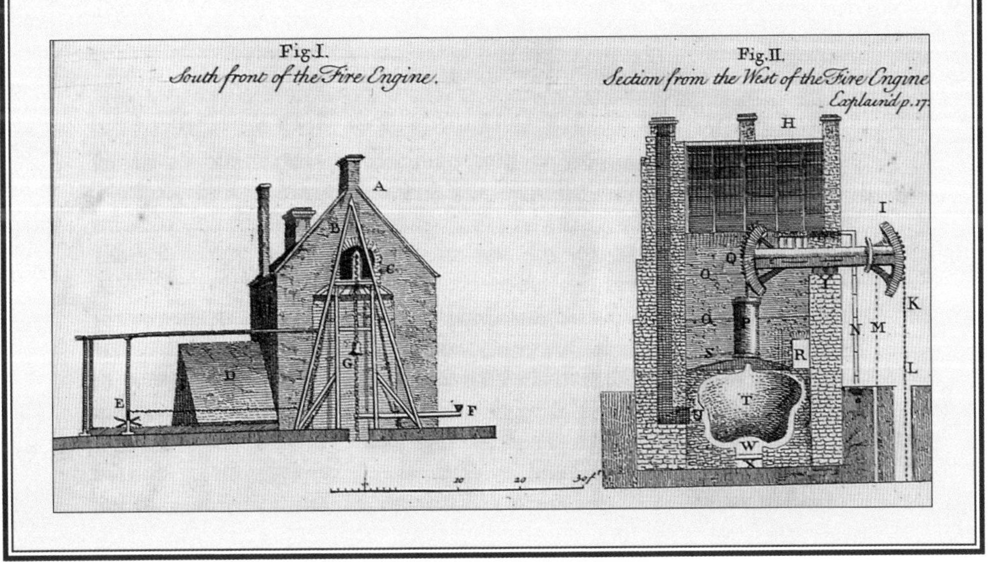

or Great Work. Within a few years several of these were pumping from the deeper mines of Cornwall, and despite the reintroduction of sea-borne coal duty by Walpole's government, which made their use cripplingly expensive, Newcomen engines enabled the copper mines to continue to operate as they became deeper. In 1741 the Cornish political lobby led by William Lemon persuaded Parliament to remit sea-borne coal duty once again, specifically to ensure the survival of the copper mines. Newcomen engines, which had languished due to the cost of coal, were restarted as new engines were purchased and the mines once again became viable. John Smeaton came to Cornwall in the middle of the 18th century and worked on improvements to Coster's water engine, as he did in the 1770s to Newcomen's atmospheric engine. He vastly improved the performance of Coster's and more than doubled the efficiency of Newcomen's, but despite this both types of engine were beginning to reach the limits of their effectiveness.[19]

Drainage adits

As noted in earlier chapters, the problem of drainage in Cornish mines was a constant one and, as in medieval, Tudor and Stuart times, pumping machinery could not provide the whole answer. With the introduction of gunpowder shothole blasting in the 17th century, drainage adits could be extended into mines at a far faster rate than hitherto, and the first half of the 18th century saw an explosion of adit driving. Not only short, local, single mine adits were driven, but regional adits too, extended to serve whole groups of mines. When the century began hundreds of adits had been driven from the nearest cliff or valley, most entering the mines less than 50 ft (15 m) from surface, and some being no longer than 100 to 150 ft (30 to 45 m). The sixteenth-century adits referred to by Richard Carew, which came across country 'athwart the lodes', were rare. Things changed with the advent of copper mining and the far greater capital available to the adventurers. One adit that was to have a far-reaching effect on attitudes to mining was Pool Adit, started in 1710. This was driven from below Tuckingmill for three-quarters of a mile into the ancient Pool mine and went below the main workings at a depth of some 120 ft (36 m), draining a large block of hitherto unworked ore ground. What made the adit special was that, in following a rich copper lode in an easterly direction, it intersected half-a-dozen other rich copper lodes. These lay on an eastnortheasterly strike. Vast new copper ore reserves were opened up and the mineral lord, Francis Basset, made a fortune. During one ten-year period in the first half of the 18th century the Bassets cleared a profit of £110,000 from Pool Adit alone. Undoubtedly inspired by this success, adits were driven for longer distances and at greater depths throughout the mining districts of Cornwall. The first half of the 18th century saw regional adits being started in every important mining parish. Gwinear Adit was driven from the River Hayle for over two miles to Relistian and Wall mines, with branches extended to a score of other mines en route. Kestell Adit was driven into several mines from the west side of the Hayle River, and Barncoose Adit was extended from Wheal Fortune and Wheal Tehidy over a mile to the south, helping to drain East and West Barncoose mines and also Wheal Druid and workings on the south flanks of Carn Brea.[20]

In 1748, in order to drain Poldice mine – which was well over 600 ft (180 m) deep and flooded for much of the year – two of the greatest mining men of the century combined to drive perhaps the most important single piece of mine engineering in Cornwall's long history: the Great County Adit. William Lemon was so important in Cornish copper mining that he was known as 'The Great' Mr Lemon. Born to a poor family in Germoe in the 1690s, he spent his early boyhood earning a few pence a day as a stamps watcher. He worked his way up from clerk to manager of various businesses in west Cornwall before gaining control of Wheal Fortune, in Ludgvan. From this rich copper mine he is said to have cleared a profit of £10,000, with which he moved to Truro and became principal adventurer in Poldice and several other copper mines in Gwennap. His manager was John Williams, whose family had been acting as mine agents and managers for the landed gentry since the beginning of the 18th century. John was one of a dynasty of Williams who became the most successful mine managers and mine owners Cornwall has produced. From humble beginnings in seventeenth-century Stithians, by 1800 they controlled a quarter of all the mines in Cornwall, going on to become wealthy landowners, members of parliament and baronets. In the 19th century they became successful copper smelters in Wales, retained control of the principal copper mines until the price crashed and then expanded their tin-mining interests until the price of that metal also collapsed at the end of the century. These two men made a two-pronged assault on Poldice's water problem. Lemon used his influence to bring in the Act of Parliament to remit coal duty so that he could use Newcomen engines economically, while Williams began to plan the most ambitious adit scheme seen in Cornwall. The Act became law in 1741 and the adit was started between Twelveheads and Bissoe in 1748. Lemon ordered five engines for his Gwennap mines, two of them for Poldice.[21]

The Poldice Deep Adit, as it was first known, was to be driven for a distance of 2 1/2 miles, going up the valley to Twelveheads, along the side valley to Haile Mills, then past Killicor and Creegbrawse to the eastern end of Poldice Bal. It would then turn in a westerly direction and go right across Poldice to the edge of St Day. It was completed in about 18 years and its cost was entirely borne by Lemon and his fellow adventurers. Once the adit reached Haile Mills a branch was started towards Wheal Virgin and Wheal Maid, and, as the main drive turned across Poldice Bal, another branch was on its way to Creegbrawse and Chacewater mines. By the 1790s branches of Poldice Deep Adit were draining all the great Gwennap mines that made up the Consols and United mines, together with several others well to the south and west. The mines of St Day were also drained, as were those across the east and north sides of Redruth and Chacewater. By the end of the century it was known as the Great Gwennap Adit, and later it came to be called the County Adit. In 1800 it was 28 miles long and drained 12 square miles. By 1880, it had been extended to an incredible 40 miles and drained 16 square miles in five parishes. More than 60 mines were drained by the adit, and such was its importance that for the 160 years following its inception few of even the greatest and most profitable mines in the Gwennap and Redruth area would have survived long without it.[22]

The engineers of the 18th century

The story of this great period of expansion and success would not be complete without reference to the many engineers who were involved in it. Some of these men moved to Cornwall from other parts of the country and some were home-grown. We

EARLY CORNISH MINERS AND MIGRATION

The migration of Cornish miners is invariably seen as a nineteenth-century phenomenon. In fact Cornish miners had been migrating to find work as far back as the early 18th century, moving both within Cornwall and to other parts of the British Isles in search of better wages and conditions. In 1731 Cornishmen reopened the Old Darren mine near Aberystwyth and by the late 18th century were active in mines in the Mendips, Somerset, in the Quantocks, Shropshire, in Wales and in the Peak District. They were also numerous in the Tamar Valley and Tavistock district, as the parish registers there attest. Many moved there from west Cornwall, including the future evangelist, Billy Bray from Twelveheads, who was then an unknown copper miner.

Cornish miners had also been travelling overseas. In 1753 Josiah Hornblower (1729–1809) erected a Newcomen engine in a copper mine in Belleville, New Jersey, the first of its kind in the Americas. He expected to return to Cornwall but was persuaded to remain, and by 1760 he was running the mine himself. John Nancarrow was another 18th-century Cornish engineer who emigrated to America, and his descendants are still there. After the British took Canada from the French in 1763, an expedition was sent in 1766 to the Ontonagon River to look for copper which the Indians said existed near Lake Superior. This resulted in the discovery of a huge, two-ton piece of metallic copper, which was dubbed the 'Ontonagon boulder'. This fabulous find led to a second expedition in the early 1770s to explore the region more thoroughly. The British government looked to the Cornish to supply the miners to carry out the exploration and exploitation of this new mineral wealth. One result was to advertise the copper reserves in the Keweenaw Peninsula, Michigan, although it was to be the middle of the 19th century before improved communication, transportation, technology and finance made the copper ore there economic to extract. Cornish miners were then to lead the opening up and mining of these enormous reserves.[23]

have noted the contribution of John Coster and his son in inventing the pumping engine powered by large overshot water-wheels, and Coster has also been credited with introducing other improvements to Cornish mining practice, particularly with respect to ore-hoisting machinery. A Cornishman, Francis Scobell, also patented an engine for raising water from mines in 1724, although how widespread its use was is not known. When Newcomen introduced his steam engine into Cornwall he was accompanied by several engineers from the Midlands who helped to erect them. Prominent among these were two engineers, a Mr Darlington and Joseph Hornblower. We know little about Darlington other than that he was an engineer who was associated with the Hornblower family and that he arrived in Cornwall from the Midlands in the 1720s. Joseph Hornblower was a devout Baptist from Staffordshire, and not only did he remain in Cornwall as Newcomen's principal engineer and engine erector, he fathered a number of sons who also became important steam engineers and remained in the county.[23]

Hornblower erected several Newcomen engines in the 1720s, including William Lemon's at Wheal Fortune, Ludgvan, and others at Wheal Rose, Wheal Busy and Polgooth mine. John Wise came to Cornwall from Warwickshire in the 1740s, when steam engines were again coming into use following the remission of sea-borne coal duty. During the period 1741 to 1778, nearly 60 Newcomen engines were either restarted or purchased new from Coalbrookdale Foundry, and John Wise and his son played an important part in this. Like Hornblower, Wise lived at Truro, and as well as being 'visiting engineer' for a number of mines, he erected engines. He was employed in the 1770s at Dolcoath mine and was associated with the mines on North Downs, Redruth.[25]

Among the Cornish engineers who rose to prominence at that time were George John, John Nancarrow and John Budge. George John was the younger son of Henry John, the founder of Dolcoath, and by the 1730s he was engineer at both Dolcoath and Pool Adit, where he looked after their large water engines. He also travelled abroad to buy iron for a foundry he set up at Tuckingmill. He became skilled in brass work and his foundry supplied engine parts for several local mines, including Dolcoath, Cooks Kitchen and Great Work. He was a competent steam engineer and supervised the erection of the Newcomen engines at Dolcoath. George John was one of the principals behind the setting up of the copper-smelting furnaces at Bullen Garden in the 1750s and an important shareholder in the Cornish Copper Company, which was

The 1741 Act of Parliament for the remission of sea-borne coal duty to Cornish mines. Pressure by the powerful Cornish mining lobby, led by 'The Great' Mr William Lemon, brought the remission about and greatly reduced the cost of coal at the mines, enabling adventurers to reintroduce the coal-hungry Newcomen atmospheric engines.

(727)

Anno decimo quarto

Georgii II. Regis.

An Act for granting to His Majeſty the Sum of One Million out of the Sinking Fund, and for applying other Sums therein mentioned, for the Service of the Year One thouſand ſeven hundred and forty one; and for allowing a Drawback of the Duties upon Coals uſed in Fire Engines for draining Tin and Copper Mines in the County of *Cornwall*; and for appropriating the Supplies granted in this Seſſion of Parliament; and for making forth Duplicates of Exchequer Bills, Lottery Tickets, and Orders, loſt, burnt, or otherwiſe deſtroyed; and for giving further Time for the Payment of Duties omitted to be paid for the Indentures and Contracts of Clerks and Apprentices.

OST gracious Sovereign: We Your Majeſty's moſt dutiful and loyal Subjects, the Commons of Great Britain in Parliament aſſembled, being deſirous not only to raiſe ſuch Supplies as are neceſſary to enable Your Majeſty to carry on the preſent War with Succeſs, but alſo to uſe ſuch Ways and Means therein, as that Your Majeſty may have the better and more ſpeedy Effect of the ſaid Supplies, have reſolved to give and grant unto Your Majeſty the Sum of

Preamble.

8 F 2

of

one of the great success stories of eighteenth-century Cornwall. John Nancarrow was engineer at Great Work mine in the 1750s and was responsible for erecting a Newcomen engine there. His son, also John, followed his father into steam engineering, eventually taking his expertise to America, where he remained. Like George John, John Budge was a Camborne man, and from an early age he worked at Dolcoath mine under the former's supervision. In 1766, when he was 33 years old, he was in charge of erecting a Newcomen engine at Dolcoath. The principal adventurer at Dolcoath wrote: 'This engineer is Mr John Budge who acquired his experience under Mr George John and he is thought a very capable person.' Despite this, it was thought necessary to seek assurances about his work 'as he never finished an engine but under the direction of Mr George John'. Budge went on to be the most respected and experienced steam engineer in Cornwall, becoming chief engineer at Dolcoath and training the great inventor, Richard Trevithick, whom he took under his wing when only 15 years old.[26]

1750–75: a thriving industry

The years between 1750 and 1775 saw the Cornish copper mining industry highly productive, well-organised and confident. Vast wealth was being created for the landowners, adventurers and entrepreneurs. The more able, skilled miners were also doing very well. Many were moving socially and financially upwards, becoming mine captains and engineers, and many ordinary working miners were enjoying quite long periods of stability in their employment. Alongside the established mines – which quietly went on making good money for their owners and workers – were the spectacular successes, like Wheal Virgin in Gwennap. A rich copper lode was discovered there at a shallow level in 1757, and within a few short weeks, at a cost of not much more than £300, some £15,300 worth of copper ore was raised and sold.[27]

The number of mines also increased steadily during the period, with some 90 selling ore at the ticketings of 1770. The industry was still dominated by about 20 large producers, with Dolcoath now the largest and most productive mine in the county. In 1754 another Swedish industrial spy, Reinhold Angerstein, visited the Cornish copper mines, and he described in detail how they worked, the machinery they used, how deep they were and what their ore contained and was worth. He commented that Dolcoath was 480 ft (156 m) deep at that time, with a Newcomen engine and three water-wheel engines. Bullen Garden mine, to the east of Dolcoath and working the same lode, was producing between 500 and 600 tons of copper ore a month, and Angerstein clearly believed that it was the most impressive mine he had visited. In 1765 these two important mines amalgamated to create the largest and most successful mine in Cornwall, employing hundreds of workers and with some of the finest engines. No other copper mine could compare to this new mine, and its fame spread throughout the world. Dolcoath and Bullen Garden were at the forefront of copper production for most of the 18th century, and they were important contributors to the enormous increase in copper production, from 6480 tons a year in the 1720s to 26,400 tons a year by the 1770s. Income from Cornish copper rose in that period from £47,350 per annum to £177,833 by 1770. Although the price paid to the mines for ore dropped on average from £7 16s a ton to £6 15s, it was still high enough for the mines to make a healthy profit. As the copper mines expanded and deepened, so tin production in most of those same mines increased. In 1715 about 1500 tons of tin metal was produced in the stannaries, but by 1775 this had doubled to some 3000 tons a year, much of it still coming from alluvial stream works. Despite this story of continuous boom, there were clouds on the horizon, and these were soon to cast a shadow over the whole Cornish copper mining industry.[28]

An extract from the 1769 accounts of Francis Basset of Tehidy. The page illustrated here shows the costs paid by Mr Basset to New Dolcoath Mine for his six 16ths share in the adventure.

Dr. The Exrs. in Trust of Francis Basset Esq. in Acct. with Jas. Tippet ... Cr.

1769 To 6/16 of £ 1448.15.9 New Dolcoath Cost for July £ 543 5 11
To 6/16 of £ 2485.14.10 ... Do ... for Augst. ... 932 3 ¾
To 6/16 of £ 2197.18.4 ... Do ... for Sepr. ... 824 4 1½
To 6/16 of £ 1239. – 5 ... Do ... for Octo. ... 464 12 8
To 6/16 of £ 1799. 8. 1 ... Do ... for Novr. ... 674 15 6½
Total £ 9170.17.5 .. 6/16 thereof ... £ 3439. 1. 6¾
To the Opposite Dr. on Child Esq. & Compa. being lost 360..

– £ 3799. 1 6¾

1769 Sepr. 8 By Ballance of an Acct. settled with Mr. Basset at Tehidy to end of June ... £ 18 12 6
July 31st By 6/16 of £ 1687. 8. 7 Nt. Pce. of New Dolcoath Ores sold this day .. £ 632.15.3½
Sepr. 4th By 6/16 of £ 1667. 16. – Do. of Do. ... 625. 8. 6
Octo. 2d By 6/16 of £ 1333.15.5 Do. of Do. ... 500. 3. 3¼
" 27th By 6/16 of £ 1580.13.6 Do. of Do. ... 592.15. ¾
Total £ 6269.13.6. 6/16th thereof ... £ 2351. 2. 6½
By 1/16 of £ 300.. – 5/8 of the remaining £ 250.. chd. to New Dolcoath Acct. in Augst. 1769 for which Mr. Willm. John Debits your Acct. with him £ 81.5..
By Do. chd. in Sepr. ... 81.5.
By Do ... in Octo. ... 81.5
By Do ... in Novr. ... 81.5 – 325..
Novr. 6th By Mr. Bassets Dr. on Robt. Child Esq. & Compa. 360..
Decr. 14th By Cash ¢ 100..
£ 3154 15 ½
Ballance in favour of J. T. ... £ 644. 6 6¼
– £ 3799. 1 6¾

Errors Excepted 15th Decr. 1769

Jams. Tippet

CHAPTER SIX

SYNDICATION AND AMALGAMATION

Late 18th century (1775–1800)

OPPOSITE
Painting by Terence Cuneo of Richard Trevithick and Nicholas Holman inspecting the newly invented high-pressure Cornish boiler made in Holman's foundry to Trevithick's design.

As so often in the long history of Cornish mining, external factors were to cause the most difficult problems during the period considered in this chapter. In 1768 a vast body of low-grade copper ore was discovered on Anglesey, North Wales, and within a decade or so the price of copper had become so depressed that all the Cornish mines were in crisis. This discovery coincided with another major problem in that many mines had now reached depths that made pumping and hoisting prohibitively expensive. In the deeper mines the Coster water engine and the Newcomen steam engine had reached the limits of their effectiveness, making the low price of copper the last straw. There were solutions to the problem of inadequate and expensive technology, but for the problem of the low copper price there appeared to be no easy or quick answer.[1]

Thomas Williams, a wealthy Welsh entrepreneur who controlled the Anglesey mines, claimed that he could operate profitably with copper at £50 a ton, whereas the Cornish adventurers needed a price closer to £80 to be viable. With an area of copper ore nearly 18 acres in extent and well over 100 ft (30 m) deep to quarry, Williams' costs were minimal. He had no drainage problems, a deep-water port nearby at Amlwch and a plentiful supply of cheap coal just across the bay in south-east Lancashire. He not only held the whip hand over the supply to the smelters, but when he thought they were paying less than they should he simply erected his own furnaces and smelted his own ore. The Cornish mine owners were in an impossible position, but they strove manfully to cope, and one thing they did have was cash from several decades of impressive profits.[2]

Mine drainage and steam-powered pumping engines

The first thing the Cornish adventurers had to deal with was the water problem. The existing engines could not pump from depths greater than 80 fathoms below adit, and many of the Camborne, Redruth and Gwennap mines were already that deep and struggling. Poldice, Dolcoath, Wheal Virgin, North Downs and Ting Tang were being forced to rework shallow levels and extend outwards in search of more ore. Their deepest levels were flooded in winter. Compounding this problem of inade-

quate pumping machinery was the sheer cost of coal. Those engines that were just about coping were consuming prohibitively large amounts of coal, and the income from sales of copper ore failed to cover the costs. By 1775 half the Newcomen engines in Cornwall were stopped and the adventurers were casting about for a solution. In 1769 James Watt had patented a steam engine which was to prove, for a time, the saviour of Cornish copper mining. His engine worked on low-pressure steam, had a separate condenser and was to prove vastly superior in efficiency and economy to the Newcomen engine, even after the latter had been modified by Smeaton. In 1776 some Cornish adventurers visited the Boulton–Watt factory at Soho in Birmingham. They also visited a colliery where a Watt engine was working, and were impressed. Boulton and Watt offered to erect one in Cornwall at their own expense to demonstrate how efficient it was and how much coal it could be save. A 52-inch cylinder engine was ordered for Ting Tang mine, but transportation delays due to its size meant that by the time it was erected in 1778 another, smaller engine had already been put to work. Thomas Wilson, who managed Wheal Busy and was to erect engines for Boulton, had installed a 30-inch engine before the original 52-inch order was completed. The Cornish adventurers watched the performance of these engines closely, and they quickly learned that the 30-inch engine at Wheal Busy was outperforming the highly efficient 72-inch engine that had worked there.[3]

HOW WATT'S ENGINE WORKED

The original Watt engine used the pressure of steam to force down a piston which, as in the Newcomen engine, pulled down a lever or bob. A separate condenser then converted the steam from the cylinder back to water, setting up a vacuum, which by an arrangement of valves was used to act on the opposite side of the piston to the live steam. The combination of steam pressure, vacuum and a separate condenser resulted in a far more effective use of the steam than the Newcomen engine was capable of. An added benefit was that the cylinder was not cooled between strokes, which wasted energy. Later Watt engines underwent various modifications to improve their performance.[4]

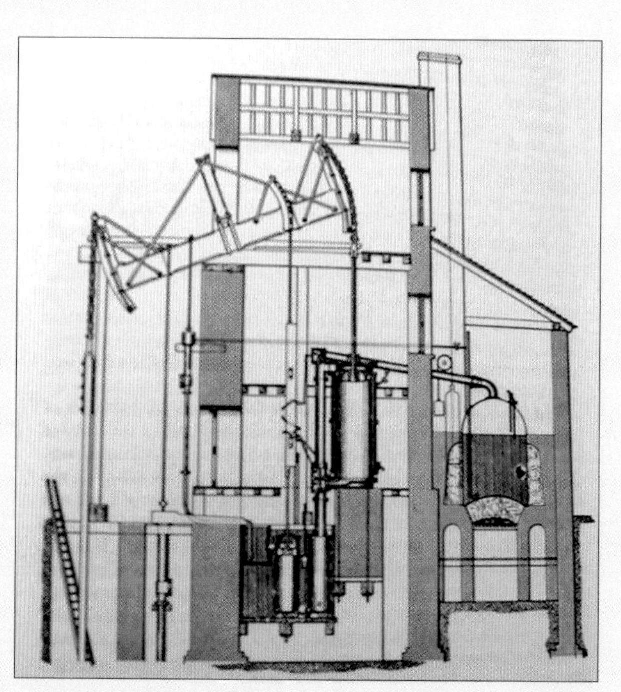

The Cornish engineers had a vested interest in the success of Newcomen engines as their livelihood depended on their installation and maintenance. Men like the older Hornblower and John Budge were not about to admit meekly that the new engines could do what theirs could not. Nevertheless, as Budge was erecting the last Newcomen to be put up in Cornwall, at Wheal Virgin, Watt's engineers were busily installing a 63-inch engine at Wheal Union mine, Goldsithney. Watt even rebuilt Smeaton's two-year-old Newcomen at Wheal Busy, installing a separate condenser and placing John Wilkinson's 63-inch cylinder inside the 72-inch casing of the original. Thereafter, with the mine adventurers convinced, Watt's engines were erected at Halaminning mine, Poldice, Wheal Chance, United Mines and Tresavean mine. In 1780 the group of mines which were shortly to form Consolidated Mines ordered five new Watt engines to replace their seven Newcomens, which were being overwhelmed by the volume of water. In 1782 Poldice purchased a second engine, and finally even Dolcoath, the greatest of the copper mines and the last to hold out against the Watt engines, succumbed and bought an engine. John Budge, Dolcoath's chief engineer and perhaps the most respected steam man in Cornwall, had been converted. By the end of 1783 some 20 Watt engines were at work in Cornwall and the last Newcomen, at Polgooth, was about to be replaced.[5]

Watt's burdensome engine patent 1775–1800

The cost of these Watt engines was borne by the adventurers of the mines in which they were installed, but Boulton and Watt built into the contracts governing the use of their engines a clause to the effect that they were to be paid one-third of the saving in coal each year. As Watt's original patent had been extended for 25 years in 1775, the mines were obliged to pay this one-third saving until the patent expired in 1800. With the price of copper metal dropping and their costs rising, mine owners had no choice but to agree to this deal, whereby Boulton and Watt were to be so handsomely recompensed for owning the patent. Despite the resentment the arrangement was to cause and the battles to which it led, the method of payment devised by the patentees was fair, although questions might be asked about the way in which Parliament sanctioned the 25-year extension of the patent.[6]

One unfortunate result of the extended patent – and Boulton's powerful and successful defence of it – was that it stifled the inventiveness and ingenuity of a whole generation of extremely competent Cornish engineers. Together with several up-country businessmen and engineers, these Cornishmen strove to create engines that could improve the capacity and efficiency of the pumping engines without falling foul of Watt's patent. In 1781 Jonathan Hornblower patented a double-cylinder engine in the form of a compound engine, with two pistons working in succession. After several years Hornblower, in association with John Winwood, a Bristol iron-

Moorstone boiler. Although traditionally attributed to Arthur Woolf senior, father of the famous engineer of that name, it appears more likely to have been originally designed and erected by Sampson Swaine at Weeth Mine, Camborne, in 1763. Woolf re-erected it at Dolcoath Mine.

founder, erected his first engine in Cornwall, at Tincroft mine. Following various trials, arguments, claims and counter-claims, they installed another nine engines in the county, and Boulton and Watt then switched their attack to the adventurers rather than Hornblower and Winwood. Eventually, these compound engines were to be used with high-pressure steam. Edward Bull, a former Boulton and Watt engineer, designed another so-called 'pirate' engine in 1790, and, with the assistance of the young Richard Trevithick, about ten of Bull's engines were installed. Bull's engine was erected over the shaft so that the cylinder worked directly on the pump and no beam was necessary. In 1793 Boulton and Watt took legal action to stop its use and the court decided that the engine was a 'manifest piracy' of the Watt patent. Following this 1794 court injunction, Bull's health declined and he died shortly thereafter. This was followed by Jabez Carter Hornblower's new engine, which was a double-cylinder rotative engine, and the Birmingham duo soon took steps to stop this one too. They were successful in the very last year of their patent – in January 1799. Despite the inhibiting effects of this litigation, men like Joel Lean, Richard Trevithick and others, together with employees of Boulton and Watt such as William Murdock, continued to experiment and modify mine machinery of all types, reintroducing the plunger pole for pumping, working on various types of hoisting engine and creating high-pressure steam engines and boilers. Murdock, Trevithick and Andrew Vivian also worked out ways to power a locomotive by steam, and Trevithick eventually made the first successful journey with his engine on Christmas Eve 1801, 'up Camborne hill'.[7]

The Cornish Metal Company

The success of Watt's engine in enabling the mines not only to pump water more efficiently and economically but also to sink deeper at a faster rate than hitherto did not solve the problem of the losses suffered by the mines due to the low price of copper. In some ways Watt's engine made the problem worse because it led to rapidly expanded output and consequent overproduction. Thomas Williams in Anglesey had no problem for the reasons stated on page 104, but the smelting companies in Bristol and Wales, known as the 'Associated Smelters', were strangling the Cornish copper industry by buying its ore cheaply. In 1785, with the copper barons beating their breasts in frustration and the miners facing hard times, Matthew Boulton and John Vivian, a leading Cornish mine adventurer, came up with a scheme to save the Cornish industry.[8]

The idea was to form a copper syndicate, which was to be known as the Cornish Metal Company. Its function would be to buy copper ore from the Cornish mines and market it itself. The principals included John Wilkinson (manufacturer), members of the Fox family (merchants), Josiah Wedgewood (manufacturer), Boulton and

Watt, Sir Francis Basset (mine owner and mineral lord) and John Vivian. Basset was to be governor of the company and Vivian was to be his deputy, carrying out the day-to-day management. The nominal capital was £500,000, of which £130,000 was subscribed at the start of business, 1 September 1785. Optimistically, Boulton declared that it was 'the most important revolution that ever happened in the mining interest … The miners and adventurers have now discovered that power ultimately rests with them and not with the copper smelting companies.'[9]

In practical terms the Cornish Metal Company was a marketing organisation. With its head office in Truro, it had a seven-year mandate to buy copper ore from the mines and sell it for as high a price as it could. It was to replace the 11 Associated Smelters, which had until then purchased all the mines' ore through ticketing, with its own marketing system. The Company would buy all the ore that was available for sale and supply it to the five smelting companies to which it was contracted. These smelters undertook to produce the copper metal in whatever form the Company required. Although there was an agreement to divide the market between the Anglesey company and the Cornish Metal Company, the inability of the latter to provide a mechanism to limit Cornish production, together with its failure to sell much of the product, led to the scheme's ultimate failure. Within months of its establishment the Company was in difficulty, from which it never managed to extricate itself. Its stockpile of unsold copper metal became an embarrassment. In June 1786 Boulton commented: 'Anglesey hath been selling and continues to sell large quantities whilst Cornwall is not selling an ounce.' Thomas Williams was not only selling copper to established customers but was creating new markets as he discovered new uses for the metal. In the autumn of 1786 James Watt wrote pessimistically: 'So great a dead stock must hang like a millstone about their necks without they take some effectual means to get quit of it.'[10]

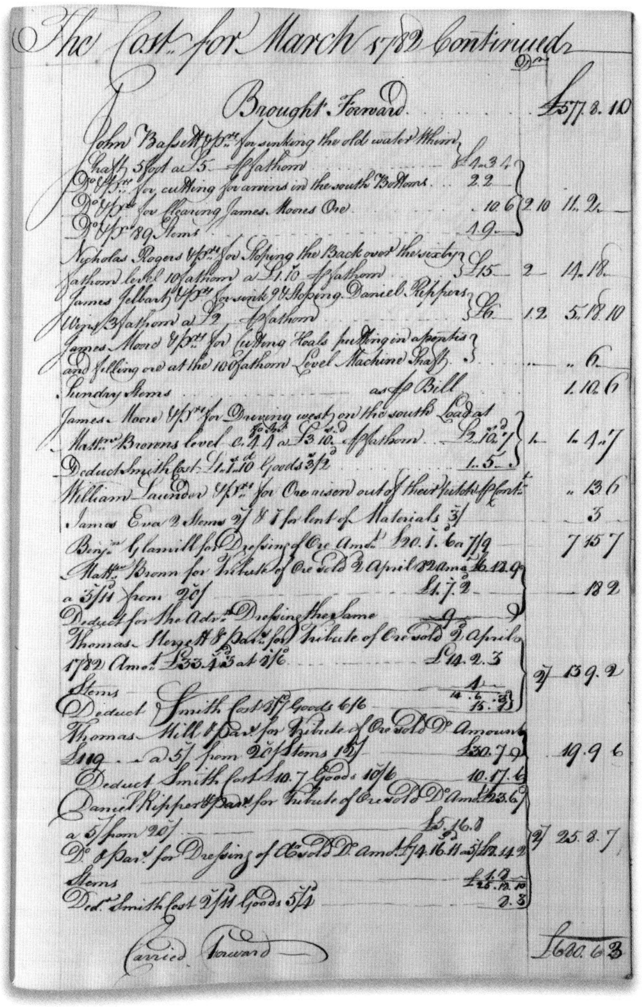

Extract from the cost book of Cooks Kitchen Mine, Illogan, for March 1782. The page has references to tutworking, tributing and men working on the owner's account. Several of the names are of families long associated with Cooks Kitchen.

In November 1787 the Cornish suffered the humiliation of having to admit failure. Due to inadequate management and with its creditworthiness gone, the Company was forced to accept terms proposed by the Welshman that appointed him sole agent for the sale of their copper for the next five years. By 1788 overproduction in Cornwall had further depressed the copper price, and by negotiation two of the largest mines, Dolcoath and North Downs, agreed to stop mining if they were paid compensation in lieu of estimated profits. In 1790 the Company relinquished to Williams its role as sole purchaser of Cornish copper ore. Williams was acting for a new association of ten smelting companies, which included his own. He had successfully schemed his way into the driving seat and now controlled almost all Welsh and Cornish copper production. What had begun as a Cornish-run marketing enterprise had been manipulated, or had evolved, into a thinly disguised smelters' combine, controlled by Williams, who very briefly held a monopoly over the whole British copper industry. By the end of 1791 Williams had not only disposed of the entire stock of the Cornish Metal Company, but he had reimbursed the subscribers their original investment and paid them all a handsome 5 per cent interest on it. He had also paid off the considerable debts which the Company had incurred during its short life. Six months before the official expiry of the Company's seven-year life, the mine owners were already going their own way and selling their ore to whichever smelter they could. It was not long before the familiar custom of ticketing resumed in the mining towns of Cornwall.[11]

Another response to low copper prices – amalgamation

One way of making mines more efficient was to follow the example of Dolcoath and amalgamate. Dolcoath had successfully done this in the 1760s, bringing together Old Dolcoath, Bullen Garden, South Entral, Stray Park, Wheal Gons and Roskear Broas. In 1780 the Gwennap copper barons amalgamated Wheal Virgin, West Wheal Virgin, Carharrack mine, Wheal Maid and others to form Consolidated Mines. This was followed two years later by Sir William Lemon and others, who created United Mines from Ale & Cakes, East and South Ale & Cakes, Cupboard mine, Wheal Moor, Wheal Chy and other small mines. North Downs, Redruth, had seen many amalgamations during the 18th century, but by the 1790s it had combined a large number of small mines into one enormous mine group. By the end of the century Poldice and Wheal Unity, long associated together, were working as one mine. During the critical decade of the 1790s these mine groups were among the most profitable in Cornwall. Although the last two decades of the 18th century were undoubtedly years of crisis, they also witnessed an impressive effort by the mine managers and miners to avert disaster and, despite the grave problems that beset them, they actually enjoyed a measure of success. Dolcoath mine, by the time of its agreed cessation of activity in 1788, had been sunk to a depth of well over 1000 ft (300m), and during the three years before closure the mine sold £62,000 worth of copper ore. The other large mine which agreed to close in 1788, North Downs, was back in production by the early 1790s and made a profit in four of the seven years between 1792 and 1798. During that period the nine largest copper mines all made

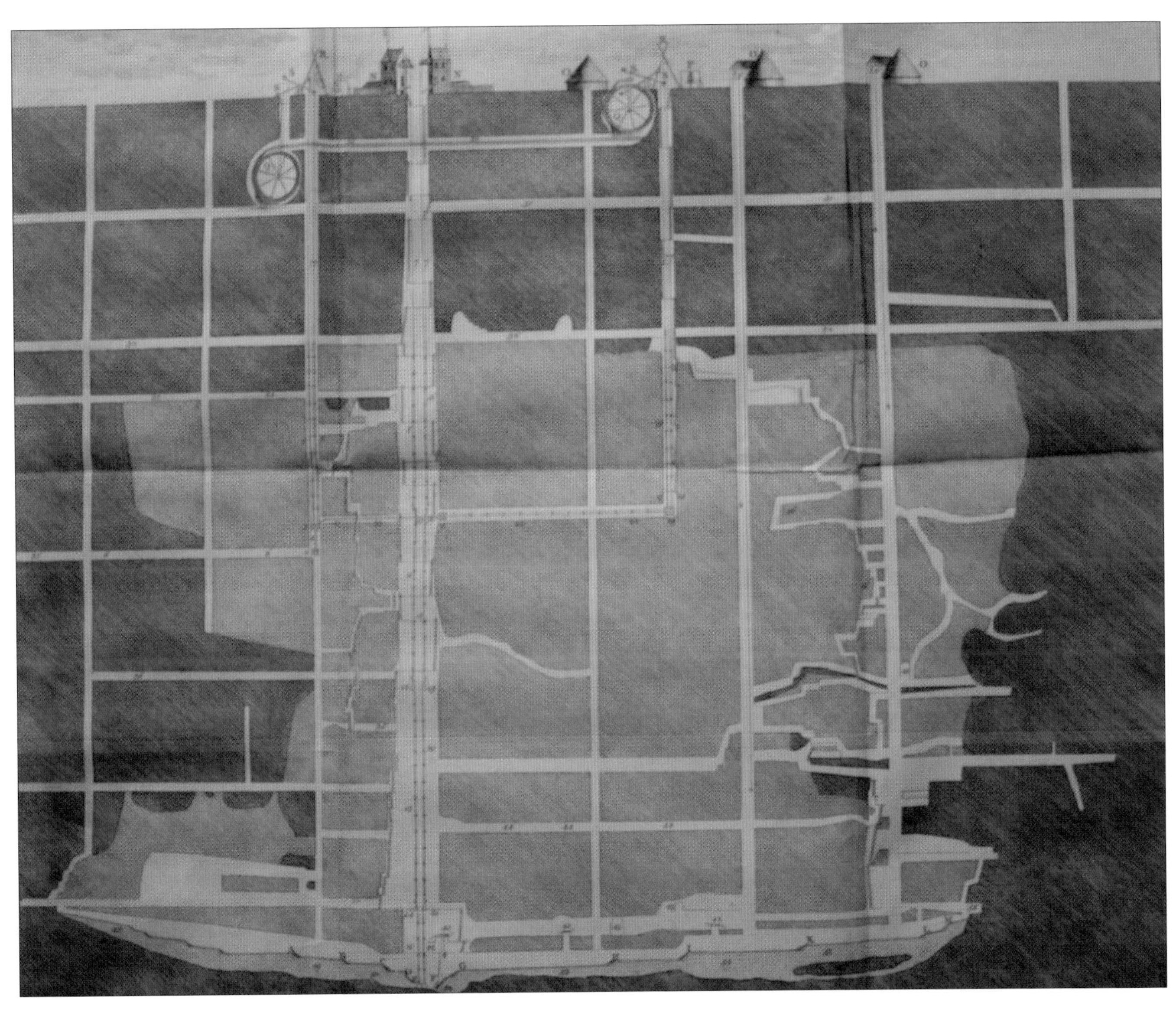

A drawing of Bullen Garden section of Dolcoath Mine. In 1765 Dolcoath and Bullen Garden amalgamated to form New Dolcoath, which became for a while the largest copper mine in the world. The two water engines and two Newcomen atmospheric engines are depicted, along with horse whims for hoisting the copper ore to surface and a dozen working levels and worked-out stopes from which a vast tonnage of copper and tin ore had been extracted. The picture was probably drawn in the late 1760s and was published by William Pryce in 1778.

profits, with Wheal Unity in Gwennap making a profit for the whole seven-year period and Tincroft, Cooks Kitchen, Crenver & Oatfield, Consolidated and United Mines all being in profit for most of that time. Overall, between the years 1792 and 1798 inclusive, £2,237,291 worth of copper ore was sold from Cornish mines, leaving a total profit for all mines of £42,168. The greatest number of copper mines at work during those years was 79, and although the majority of small mines made annual losses, the largest and best equipped mines worked profitably.[12]

In 1796 Charles Hatchett, who had travelled across Europe and Great Britain studying mining and industry, visited the mines at Camborne and Illogan, and in his diary he recorded what he saw there. The four mines he inspected lie in line along the same lode systems, with Wheal Gons on the west and Tincroft on the east of the line. Wheal Gons was 150 fathoms or 900 ft (273 m) deep and was pumped by a Watt steam engine and two 36-ft (11-m) diameter water-wheel engines, one above ground and one below. Eleven levels were worked underground and 150 men and boys were employed. The mine had four sets of stamping mills. Between 80 and 100 tons of copper ore were produced each month, and the ore was sold for an average of £9 a ton. Hatchett's guide was Captain Andrew Vivian, one of a large family of Camborne mine engineers and adventurers. Hatchett observed that Dolcoath mine, formerly the largest in Cornwall, was 174 fathoms or 1044 ft (316 m) deep. Going east, the next mine was Cooks Kitchen, which was 870 ft (263 m) deep and was pumped by means of a 36-inch Watt engine and three enormous water engines, the two engines above ground having wheels 42 and 48 ft (12.7 and 14.5 m) in diameter and the one underground having a wheel that was 54 ft (16 m) in diameter. Four shafts were used for hoisting the ore and between 300 and 350 tons of ore were produced every month, with an average value of £8 a ton. There were 340 men and boys

JOHN WESLEY AND METHODISM IN CORNWALL

Dissent in religion was not new in 18th-century Cornwall. The Cornish rebelled in 1549 against the introduction of a new prayer book, and the non-conformist Anabaptists and Quakers had also been influential among some of the better-off in Cornwall since the 17th century, but when John and Charles Wesley preached Methodism in Cornwall in 1743 the whole of the mining population was affected. At first local opinion was polarised, with large crowds gathering to listen to their preaching, while drunken mobs were encouraged to riot and attack them. The Wesleys were chased out of some parishes. Despite the opposition of the established church and its churchwardens the movement grew, mostly due to the simple message the Methodists preached: immediate salvation based on faith alone.

By 1750 there were some 30 Methodist 'societies' or groups in the western mining districts and several more in the farming parishes of the east. At first they met in cottages and barns or outside in amphitheatres like Gwennap Pit, but the societies gradually became organised into proper structures and built small, simple chapels to worship in. By 1798 the Redruth Circuit alone was the fourth largest in the country and Methodism had become a part of the Cornish landscape. The miners especially took to the simple, uncomplicated system of worship and belief. Living in poverty, with sudden death at work a daily reality, they recognised a religion which encouraged an outlook that gave their lives meaning, purpose and hope.

Much stress was placed on literacy, and local 'dames schools' and Sunday schools sprang up, enabling the children of the miners to learn to read and write. As the 19th century progressed most of the small, basic chapels were replaced by grand buildings with beautiful timberwork and organs, reflecting the growing wealth of the church. In mining districts it was normal for the mine captain who directed labour underground to preach the sermon in chapel on Sunday. By 1851 Cornwall was the only county in the country where the majority of the population claimed to be Methodist. There were many offshoots of the original movement, but they all sprang from the same ideas and eventually all became part of the present Methodist Union.[14]

employed, with 170 underground. During the previous ten years Cooks Kitchen had made a profit of £100,000. Tincroft was the most easterly of the mines to which Vivian took Hatchett, and it was 486 ft (147 m) deep in 1796. The Hornblower engine was still working on the mine, which employed 200 men, 140 underground. All these mines were profitable in the 1790s, and although 690 employees was not a small workforce, when compared to the many hundred Dolcoath had employed a short time before, it left many out-of-work miners in and around Camborne.[13]

Revived importance of tin

Although eighteenth-century Cornwall is considered to be all about copper, tin mining and streaming continued and even expanded. In 1700 annual Cornish tin production stood at just over 1400 tons, but by the 1790s the annual average had risen to 3245 tons. Thus, as the copper crisis bit deeper, income from tin became a crucial factor in many a mine's survival. Old workings, mostly above the copper levels, were reworked for tin, and old skills had to be relearned by the tributers as they were forced to identify and evaluate an ore which was harder to recognise than the more obvious copper ore. An example of this trend towards an interest in tin comes from Dolcoath, which had not been a big tin producer since the early part of the 18th century. In January 1788 the local stannary court had to decide a case between two groups of tributers, one of which was working in a stope at the 80-fathom level, and the other which had taken a section called 'Riddler's pitch'. Each group of tributers was to be paid 10s in the pound for the tin raised, and at the end of the three-month period they would divide their takings equally. However, Henry Sampson and his pare had better ground than Richard Luke's pare and refused to honour the agreement. Luke claimed he was owed £21 or more. This tells us that tin tribute bargains were being set at Dolcoath, even at the lower levels of the mine, during her last desperate attempts to keep going in the face of mounting losses.[15]

A major problem which the tin proprietors increasingly faced towards the end of the 18th century was the low price they were paid by those who controlled the market: the pewterers. To remedy this they decided to take the sale of tin metal into their own hands. In cooperation with the Vice Warden of the Stannaries, Henry Rosewarne, two respected mining men, Pascoe Grenfell Senior and John Vivian, helped to organise an Association of the Proprietors of Tin, which held its first meeting on 28 January 1780 in Helston. Its stated purpose was 'to rescue the trade out of the hands of the pewterers, who for many years past have acted diametrically opposite to the true interest of our county'. The Association appointed five agents to sell its tin in London and three to sell it in Bristol. They held regular meetings in Truro, Penzance and Helston, mostly to decide the minimum price per hundredweight at which tin was to be sold. From £3 10s (£70 a ton) in 1780, the price was gradually raised through their efforts to £4 18s (£98 a ton) by 1800. The Association remained an important factor in the Cornish tin industry until the abolition of coinage in 1838.[16]

Poverty and social disorder

A problem that persisted throughout the 18th century – despite the success of mining and long periods of full employment in some districts – was poverty. A symptom of the widespread poverty was violence, with wrecking and food riots a feature of both countryside and town. Although wrecking had long been a source of random and occasional windfalls for the poor, it was during the 18th century that, alongside smuggling, it became a regular source of income for all those who worked or lived close to the Cornish coast. Even the gentry, many of whom claimed ancient wreckers' rights, participated in the business. They might not rise at midnight and climb down the cliffs for loot, but when day came, supported by armed retainers, they would seek to drive off the mob – who might have been struggling all night in the surf – to claim what they regarded as God's gift to them. Dr William Borlase's brother, George, gave a chilling description of the tinners at work when a wreck was washed ashore. If they saw a ship in trouble off the coast the miners would leave their work, sharpen their axes and track the vessel to the cove where it was finally wrecked. He wrote that it was seldom fewer than 2000 miners who descended on the unfortunate ship's crew, looting everything they could lay their hands on, and woe betide any half-dead sailor who tried to stop them. These men could 'cut a large trading vessell to pieces in one tide and cut down everybody that offers to oppose them', wrote Borlase. The coast between Porthleven and Marazion was especially notorious, and a saying from the time expressed the sailor's fear:

> God keep us from shelving sands
> And save us from Breage and Germoe men's hands.

This stretch of coast was probably no worse than any other, but it gained a reputation for lawlessness and violence that, even in riotous eighteenth-century England, was considered unusually bad. The poor were also involved in smuggling, and again miners were willing participants, especially when work was short or non-existent. Harry and John Carter, at Prussia Cove, became infamous as 'free traders', bringing contraband from France and fighting off the revenue officers when necessary. John Carter's flamboyant life-style earned him the nickname 'The King of Prussia'.[17]

Much more common in the 18th century were food riots. Unemployment, starvation, harvest failure and high corn prices were the usual causes of such riots, and they occurred throughout England in that century. In 1727, when corn was scarce, severe rioting and looting took place in west Cornwall, and only the intervention of Sir John St Aubyn – who provided money to buy corn for the poor – saved the situation. The rioting and violence that broke out all over Cornwall due to food shortages in 1729 became so severe that the ringleaders were hanged. In 1748, with rumours that corn was to be exported from Penryn, a large mob of miners and others descended on the port and was only brought under control when troops were sent from Falmouth. Two looters were shot dead by the soldiers before order was restored. In 1757 at Padstow, in 1766 at Truro and Redruth and in 1773 again at Padstow, miners rioted and looted much-needed and prohibitively expensive corn. In 1773, when the looters moved from Padstow to Wadebridge, the army was again called out, and the unfortunate soldiers – 16 in number – sent to quell the riot were disarmed by the miners. Despite the general lawlessness of England in the 18th century, Cornish miners were viewed as especially dangerous and given to riot and violence. Even John Wesley commented, in 1745, on the bad behaviour of the miners of Gwennap who, 'made drunk on purpose, were coming to do terrible things'. Wesley failed to calm them and his peaceful meeting broke up.[18]

Towards the end of the century, when the crisis created by the falling copper price struck the miners, they turned their hatred away from the corn factors and exporters to other targets. There were still riots over food shortages and high prices, as in Truro in 1789, when unemployed miners rioted and looted – but they had also identified another kind of culprit: Boulton and Watt. Their coal-saving levy was blamed for the predicament of the struggling mines, and riots at Poldice in 1787 and 1789 were specifically directed at them. Another target of the miners' anger was the Cornish Metal Company, which was perceived as having exacerbated the situation it was meant to improve. James Watt blamed the mine adventurers for stirring up the miners against them. He wrote: 'I am very much alarmed at this fresh rising of the miners, who are certainly instigated by some enemy to the county … I hope no improper concessions will be made to them, and that some body of authority will interpose in time, soldiers should be quartered at Truro and Redruth.' Nastily, he later moaned that the soldiers had refused to fire on the rioting miners: 'I hope it is not so, otherwise we are going the same road as the French are.' Fortunately for the miners, many adventurers took a different view, and money and support were provided by them to the poor.[19]

Sir Francis Buller Bart (1764-1802). The Bullers were significant mineral lords and mine adventurers from at least the 17th century, making a fortune from both tin and copper.

In 1795 a very serious situation arose in which widespread violence and mob action were apparently being tolerated by the authorities, who seemed to be helpless in the face of the crisis. The rioters were in charge, and they forced the millers

Painting by John Opie of Thomas Daniel and Captain Morcom. The miner is pointing toward an engine house and Thomas Daniel is holding a mineral specimen. Late 18th century.

and dealers in grain to sell at the price the mob demanded. Sir Francis Basset, who was in London when he heard of the magistrates' impotence, returned immediately to Tehidy where he interviewed the millers and corn merchants to ascertain the true position. Swearing in 80 constables, he moved with great speed against the leading rioters, taking 50 of them in their beds at two o'clock in the morning. They were carted off to Bodmin gaol to be tried at the next assizes. Three were condemned to death and others to transportation to Australia. As was the custom of the time, the judge consulted with Basset and they agreed to hang one man as an example and pardon the other two. A vast mob of angry people met the body of the hanged miner, one Hosking from Troon, a village near Camborne, and things looked ugly for a time, but eventually it was accepted that in the circumstances the authorities had shown some leniency and the situation quietened down.[20]

Despite the endemic lawlessness of the period, Cornish miners were patriotic, seeing themselves as loyal Englishmen. The Navy had large numbers of Cornish sailors of every rank, from the lowliest gunner to the most important admiral, and many of England's most famous sea captains were from Cornwall: men like Admirals

Boscawen and Pellew, and Captain Bligh of the Bounty. When danger threatened, miners were quick to offer their services in the country's defence. When Jacobites raised the flag of rebellion in 1745, the 'Independent Company of the Stout, True, and Hearty Tinners within and belonging to our said parish of Redruth' volunteered to assist King George against Bonny Prince Charlie. In 1779, when a Franco-Spanish fleet threatened invasion, the miners of Dolcoath volunteered to march to Plymouth to help defend it. They remained on standby until 1791. In 1794, with another threat of invasion, a company was raised that called itself the 'Royal Redruth Infantry', and again in 1803, when war against France and her allies resumed, the miners of Pednandrea and New Wheal Virgin offered to serve under their mine captains to defend the country. A report to Parliament in 1806 shows that companies of miners in all the Cornish mining districts were organised in the country's defence.[21]

A new dawn

The end of the 18th century saw Cornish mine owners and miners breath a collective sigh of relief as the vast copper orebody on Parys Mountain, Anglesey, was finally exhausted above the 120-ft level, where it was replaced by a series of narrow copper-bearing veins. Within a few short years the price of copper metal had doubled, and with the Navy's increasing need for the metal to sheath her copper-bottomed warships it did not look as though the demand would reduce in the foreseeable future. At the same time Watt's onerous patent on his engines finally ran out, and although several mines still owed Boulton and Watt large sums of money, it would be left to the latters' sons to collect it when they could. The ingenuity and skills of the many Cornish mine and steam engineers were now to be given free rein, inaugurating a period of almost unprecedented inventiveness and innovation. Richard Trevithick, William West, Samuel Grose, Andrew and John Vivian, Richard Jeffree, Matthew and Michael Loam, Joel Lean, James Gribble, Charles Woolf, Nicholas Holman, William and James Sims, Henry and John Harvey and a host of others were to transform the worlds of mining and engineering, inventing, innovating, designing and building in ways that were to help Britain to become the manufacturing centre of the world and the greatest trading nation in history.

CHAPTER SEVEN

CORNISH COPPER BACK ON TOP

19th century (1800–66)

OPPOSITE
Map of Chiverton Mining District, Perranzabuloe, dated 1869, by E. W. Brunton. This regional mining map shows the large number of mines in the district, together with the scores of lodes worked, the adits draining the mines and the principal shafts. It is based on the 1840 Tithe Apportionment Maps for the area.

FOR THE CORNISH COPPER MINING INDUSTRY the year 1800 represented much more than the start of a new century. It marked the end of the struggle against cheap copper from Anglesey, the end of Watt's patent, the beginning of a period of tremendous creativity on the part of Cornish engineers, and, although it was slow in getting under way, the start of the greatest and most productive period of Cornish mining. The early years of the 19th century also saw moves by Cornish mine owners and entrepreneurs to eradicate once and for all the dominance of their industry by smelters in Wales and Gloucestershire. The great Cornish mining families took control of their own destinies during these years by moving their money into Welsh smelting, of which they were to retain control for most of the century.[1]

Cornish copper leads the world – the early 19th century

By 1799 the price of copper metal had risen to £120 a ton, which was twice what it had been in 1790. Captain Andrew Vivian, with Richard Trevithick as his engineer, approached Lord de Dunstanville for the lease of Dolcoath mine, and within a few months was on the way to reopening the greatest of Cornish copper mines. Backed by the Basset, Fox and Williams families, the old mine quickly regained its position of dominance in the copper ticketing. By 1805 the price of the metal had risen to £138 a ton, the highest it had been, and other long-closed mines began to reopen. Discoveries of ore made during the decades of crisis meant that new enterprises could be set up, several of which were very profitable. Once again Cornwall became the world's leading copper producer, and it remained so for several decades. Dolcoath remained the top copper producer until 1816, when other mines came to the fore – some suddenly and spectacularly, like Crinnis at St Austell, and others like the great Gwennap mines for much longer periods. In 1824 John Rule, the manager of Dolcoath, reluctantly relinquished his traditional position at the head of the ticketings to Taylor's Consolidated (Consols) and United Mines.[2]

An increased demand for copper, due largely to the Napoleonic Wars, drew capital and workers into Cornwall. The 'Wise Men of the East', as the Cornish derisively

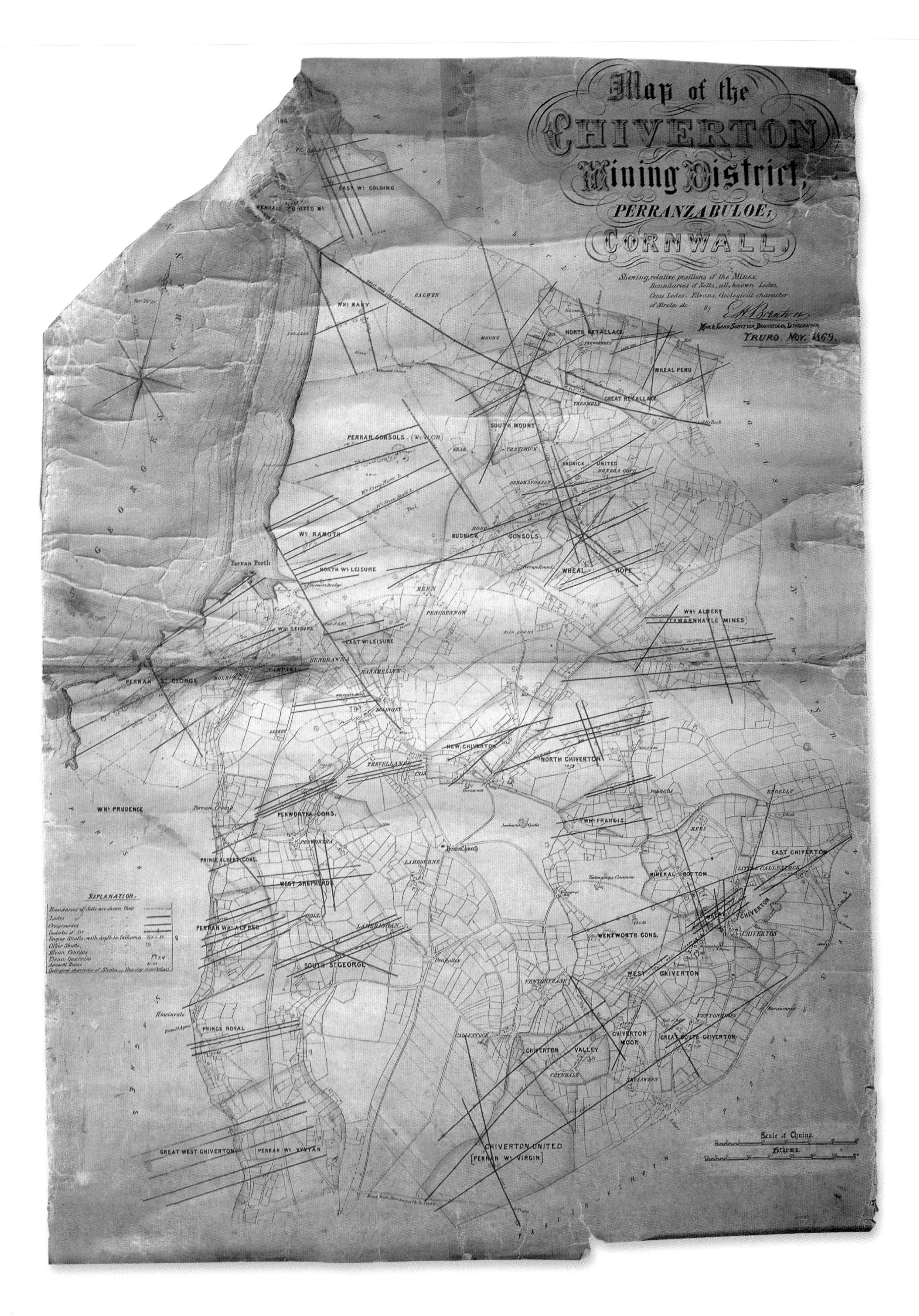

Map of the
CHIVERTON
Mining District,
PERRANZABULOE;
CORNWALL.
Shewing, relative positions of the Mines, Boundaries of Setts, all, known Lodes, Cross Lodes, Elvans, Geological character of Strata &c.
By E. H. Brenton
Mine & Land Surveyor, Draughtsman, Lithographer,
TRURO, NOV. 1869.
EAST Wh GOLDING
PENHALE UNITED Ms
Whl MARY
HALWYN
MOUNT
NORTH RETALLACK
WHEAL PERU
GREAT RETALLACK
SOUTH MOUNT
PERRAN CONSOLS
TREVITHICK
BUDNICK UNITED
HENDRA COOM
HENDRAVOSSAN
BUDNICK CONSOLS
Whl RAMOTH
NORTH Whl LEISURE
Perran Porth
WHEAL HOPE
BEEN
PENCRENNOW
Whl ALBERT
TYWARNHAYLE MINES
Whl LEISURE
EAST Whl LEISURE
HENDRAWNA
NANSMELLYN
PERRAN St GEORGE
BOLINGEY
NEW CHIVERTON
NORTH CHIVERTON
TREVELLANCE CON
Whl PRUDENCE
PENWORTHA CONS.
Whl FRANCIS
PRINCE ALBERT CONS.
Perran Church
LAMBOURNE
REES
EAST CHIVERTON
LITTLE CALLESTOCK
MINERAL BOTTOM
WEST SHEPHERDS
EXPLANATION
PERRAN Whl ALFRED
LAMBRIGGAN
SOUTH St GEORGE
CHIVERTON
WENTWORTH CONS.
WEST CHIVERTON
PRINCE ROYAL
CALLESTOCK
CHIVERTON VALLEY
CHIVERTON MOOR
GREAT SOUTH CHIVERTON
CHYNHALE
GREAT WEST CHIVERTON
PERRAN Whl VIVYAN
CHIVERTON UNITED
(PERRAN Whl VIRGIN)
Scale of Chains.
Fathoms.

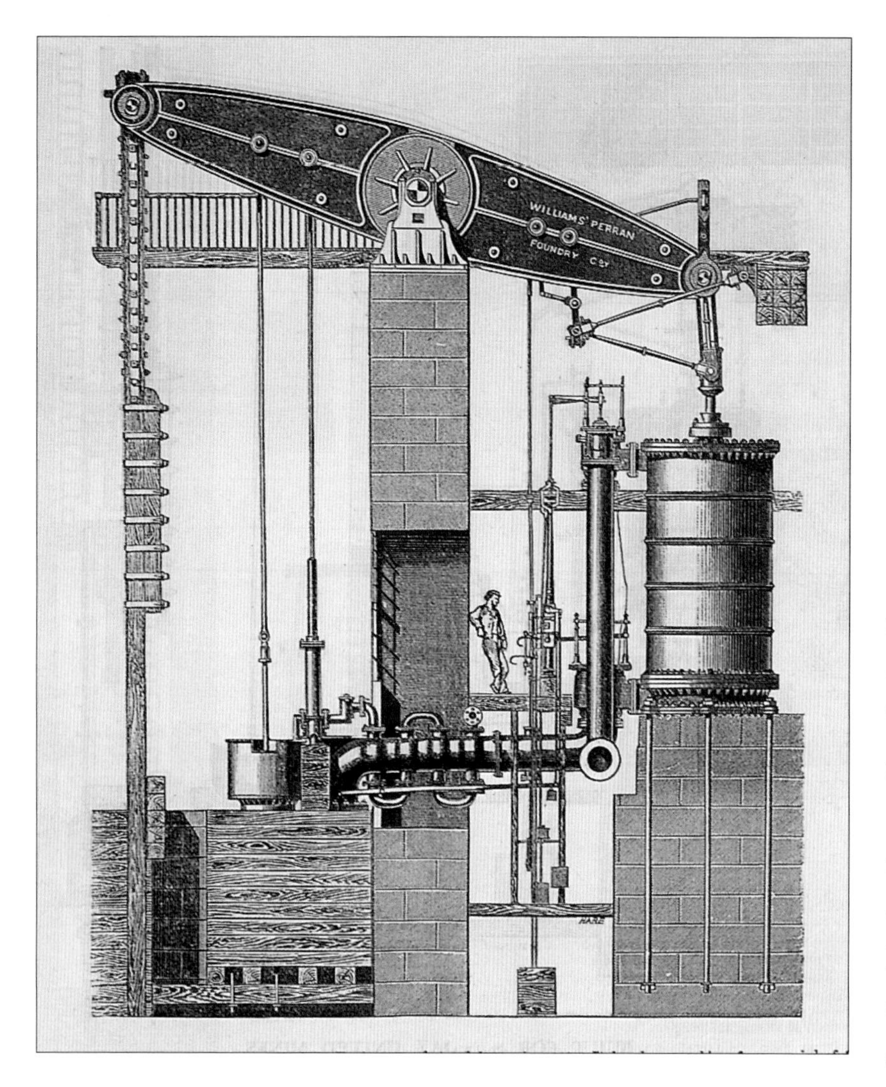

Taylor's 85-inch cylinder pumping engine. This highly efficient Cornish engine was located at Ale & Cakes section of United Mines, Gwennap. The engraving is based on a drawing dated 1840.

called these outsiders, included many of exceptional ability, like John Taylor, who came down to the west country in 1796 to run Devon's Wheal Friendship mine and turned it into a model of efficient, large-scale modern mining. Less than 20 years of age when he arrived, by his thirties he had moved westwards to Gwennap, and in 1819 he successfully negotiated leases of the extensive Consolidated Mines. In 1824 he added United Mines to his organisation, bringing into being the largest copper-producing group of mines in the world. Taylor is credited with introducing Cornish roller-crushers into his mines in 1806, and he was also one of the first to make extensive use of the steam whims created by Trevithick and Watt. His large pumping engines, some of them designed by Arthur Woolf, whom he appointed chief engineer to his mines when he reopened Consols, were among the finest and most efficient in Cornwall. Taylor's management skills were impressive, and with Captain William Davey in place to run the mining side and Arthur Woolf there to superintend the engineering, Consols and United were a tremendous success. John Taylor had outmanoeuvred the Williams family in obtaining the highly desirable Consols property and in gaining United, but Taylor never totally matched them in their far-sighted control of the largest part of the Cornish copper mining industry. The 1799 Parliamentary Report on the copper trade shows that John Williams managed 22 copper mines in Cornwall, one-third of the total, and in 1824 *Pigot's Directory* tells us that the Williams family still managed one fifth of all Cornish tin and copper mines. The Williams family remained the most important in Cornish mining for the whole of the 19th century.[3]

Once Taylor combined Consols with United their joint output was enormous, producing 20,000 tons of ore a year and employing over 3000 workers. The massive output of Crinnis near St Austell in 1816, when she produced 40,000 tons of copper ore from within 300 ft (90 m) of the surface, encouraged miners to look further to the east of the county for new sources of supply. Discoveries at East Crinnis, Pembroke and other mines in the St Austell area soon followed, and the 18th-century workings at Wheals Treasure, Fortune, Commerce, Chance, Regent and Cuddra were all re-examined. The 1820s and 1830s saw considerable exploration activity in the area, which eventually led to the opening of one of the greatest mines in 19th-century Cornwall: Fowey Consols. By 1837 this mine was the second largest copper mine in Cornwall, and for a while it was to be the deepest and most modern mine in the county. For a short period in the 1840s the owner of Fowey Consols, Joseph Austen, alias Treffry, was the biggest mine owner and the largest employer of labour in Cornwall. He built railways and canals, as well as leats for his water engines, and constructed Par harbour to bring in mine supplies and take away his copper ore. He also owned Par Consols, which after 1840 became a very significant copper producer. Much of the emphasis of Cornish copper mining had, by the middle of the century, moved east to Carlyon Bay, and it was to continue to move further in that direction as mines opened and reopened at Caradon, Kitt Hill and other areas well to the east of St Austell.[4]

Technical advances – pumping engines

Improvements to the design of steam pumping engines, pitwork and boilers followed quickly after Watt's restrictive patent expired. Trevithick's high-pressure engines and boilers revolutionised steam technology, and his non-condensing 'puffer engines' solved the problem of expensive and inefficient winding. Woolf's arrival back in Cornwall in 1812 also contributed to improved standards of engineering in the design, erection and maintenance of engines. Joel Lean's introduction in 1811 of *Engine Reporters* – a monthly publication which reported the duty (a measure of performance; see below) of engines on Cornish mines – acted as an impetus to improve performance by creating rivalry among engineers. It was not only the basic design of engines that improved in this period, for many peripheral modifications were introduced that also enhanced performance: the size and depth of the boiler furnace, better cleaning techniques and more careful lagging and insulation of cylinders. Engineers such as Jeffery and Gribble of Dolcoath advertised their skill in improving engine performance by watching them work and adjusting, modifying and altering the engine, boiler or pitwork. The service cost only £150 and resulted in remarkable savings in fuel costs as well as improved efficiency.[5]

Arthur Woolf, Richard Trevithick, William West, Joel Lean and many other engineers were designing, testing and modifying engines of all sorts and sizes. As the century progressed the duty or performance of these engines grew yearly. Lean's *Reporters* charted this progress. The duty of Cornish engines was measured in foot-pounds (ft-lb): that is, the measure of performance was how many pounds of water an engine could lift through one foot for the consumption of one bushel of coal.

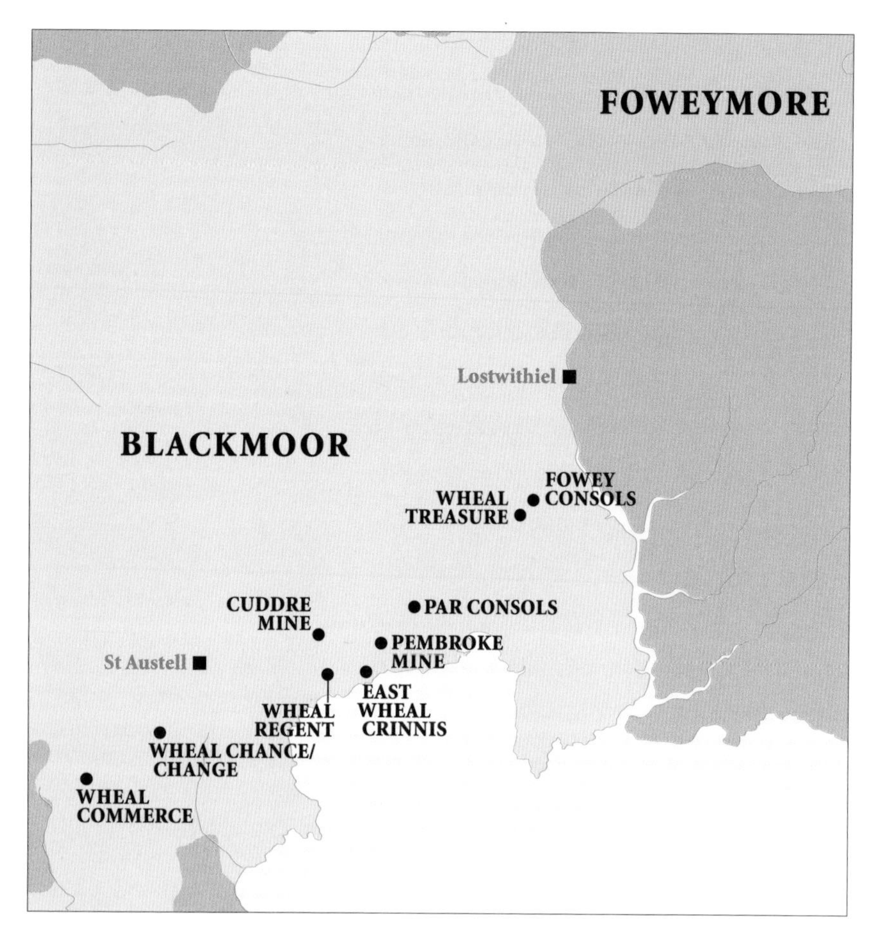

The mines around St Austell and St Blazey. During the first four decades of the 19th century these mines assumed great importance as they began to outperfrom many in the older copper-mining parishes in west Cornwall.

Unfortunately, a bushel was a measurement of volume, not of weight, which meant that the weight of a bushel of coal varied greatly. It also led to accidental or even deliberate manipulation of the figures, sometimes to enhance the claims of a particular engineer for his engine. Eventually, coal was measured by the hundredweight (cwt) of 112 lb. Duties were in the region of millions of foot-pounds. By this method of measuring, back in 1720 Newcomen's engines had given a duty of about 3,800,00 ft-lb, which Smeaton's improvements increased to nearly 8,000,000 ft-lb. Watt's original engines produced duties of about 25,000,000 ft-lb, and by the end of the 18th century the figure was rising steadily. Thereafter, following a short hiatus until about 1811, the duty figures rose constantly, until by the 1840s the best engines were achieving duties of over 100,000,000 ft-lb and averaging 60,000,000 ft-lb. Then, for a variety of reasons, the duties achieved by the best engines fell dramatically, although the average declined more slowly.[7]

Cornwall wins control of the smelters

After the removal of Anglesey copper as a threat to Cornish domination and with the increasing price of copper metal the great Cornish copper proprietors went on the offensive. The industry was still in the power of the Welsh smelters, mostly centred around Swansea, and the Vivians, Grenfells, Foxes and Williams decided that if they could not beat the smelters they would join them – or take them over. Starting with Pascoe Grenfell, in 1803, who helped set up the Copper Bank Smelting Works on the River Tawe at Swansea, Cornish money gradually took over firm after firm. By the 1830s the Williamses, Daniells, Vivians and other Cornish families controlled a large part of the smelting industry and were well on the way to dominating it. They were in such a strong position as the century progressed that, when copper ore from foreign sources began to rival Cornish production, these wealthy Cornishmen were able to retain their power and profits. Their success during the first half of the 19th century was as impressive an example of initiative and enterprise as was that of William Lemon and John Williams in the early 18th century.[8]

The waning of the stannary system

During the decades before the 1830s the stannary system – which was still in place to govern the tin industry – saw many changes. What had seemed a traditional, easy-going and almost timeless regime found itself subject to a world in which economic realities were changing constantly and rapidly. The system was buckling under the irresistible pressure to modernise. In 1837, after many years of talking about it, copper mining was also placed under the jurisdiction of the stannary courts. However, with the abolition of tin coinage in 1838 and the gradual reduction of all effective independence, the power of the stannaries was being so severely curtailed as to render them almost totally ineffective. Ironically, this coincided with the realisation among many mine captains that tin might in fact have a longer-term future than copper. For deep beneath the rich copper zones of Tincroft, Carn Brea and Dolcoath it had been noted as early as 1822 that tin was replacing copper in the main lodes. In the deeper levels of Dolcoath this trend had become pronounced by the end of the 1830s and, despite the scepticism of 'experts' from the great Gwennap copper mines, Captains Rule, Tregonning, Petherick and Thomas began to see where the future of their mine lay. The decades that followed witnessed deep tin mining outstripping copper in many mines of the Central Mining District, and from the 1860s tin, not copper, was the more modern, technically advanced and well-organised industry.[9]

West Wheal Frances, Illogan, in the 1890s. A ragging frame for tin dressing occupies the centre of the picture. Several boys have bare feet and one of the bal maidens can be seen wearing the traditional 'gook' headgear. Large numbers of bal maids worked in the tin mines throughout the final decades of the 19th century.

CORNWALL'S FOUNDRIES

The Newcomen and Watt steam engines installed on Cornish mines in the 18th century were manufactured almost exclusively in the Midlands. Before 1800 few if any engines were entirely made in Cornwall, and although some parts were made here, the casting of large cylinders lay outside the capacity of Cornish foundries. But by the end of the 18th century foundries were being established in Cornwall that would eventually be able to carry out the most exacting and complicated casting and machining themselves. After 1800 things changed rapidly, and foundries established by Henry Harvey at Hayle and by George Fox at Perran Wharf soon became proficient in all aspects of engine-making. In 1820 these two foundries were joined by another, established by the Cornish Copper Company, which until then had been for the most part an association of merchants and copper smelters. By the 1830s the three foundries were carrying out work to the highest standard, and by 1840 they were reputed to be the best in the country. For 20 years in the middle of the 19th century these three Cornish foundries, supported by many smaller and less well-known foundries, dominated the world's production of large steam engines. Almost every one of more than 600 steam engines operating in Cornish mines in the 1860s was built by local foundries. They designed and built engines from the tiniest capstan engine to the largest pumping engine. Not only did they make engines for mines across the globe, they also built engines for water works and massive engines to drain the Dutch meres. One cylinder cast by Harvey's for such an engine was 12 ft (3.6 m) in diameter. Although most engine-building was carried out by the three largest foundries, some engines were constructed by Holman Brothers and Tuckingmill Foundry of Camborne. Foundries were also established at St Austell, St Blazey, Charlestown, Penzance and Tregaseal Valley, all of which carried out engineering work for the scores of mines in their areas.

Harvey's of Hayle eventually went into ship-building and employed some 1200 workers. Its plant, which included two slipways, a forge and smithy and four steam hammers, was designed for the construction of ships of up to 4000 tons burthen. The foundry, boilerworks and ship-building yard covered 50,000 square feet (4645 square metres). Holman Brothers also diversified, becoming one of the most successful manufacturers of compressed-air machinery for mining in the country. Its compressors and rock drills were used in every mining camp in the world. It closed down only recently after more than 200 years of operation.[9]

Dolcoath copper mine in 1831. This engraving by T. Allom shows men, boys and bal maids carrying out a number of tasks at the shaft head. Bal maids are spalling, cobbing and bucking the copper ore to prepare it for sampling. *See pages 138-9 for details.*

Technical advances – the introduction of man engines

A piece of machinery that has captured the imagination of those interested in mining history is the 'man engine'. For centuries Cornish miners climbed to and from their work places on ladders. As depths increased the ladders got longer, and by the 1830s several mines were past the 1300-ft (394-m) mark and deepening rapidly. As the mines went deeper they became hotter and the problem of ventilation, never an easy one to solve, got worse. In some mines, especially Tresavean, Cooks Kitchen and United, extremely hot lodes were encountered at depth, and temperatures in excess of 100°F (38°C) were common. Older, experienced miners found it necessary to take less lucrative pitches at shallower depths, which meant financial loss to them and to the mine. Accidents in which exhausted miners fell from ladders became more common and the general health of the miners deteriorated.[10]

Thus, it was not merely for economic considerations that mine owners and managers cast about for a remedy to the problem, but also for humanitarian reasons. In the 1830s German mines in the Hartz Mountains had introduced a machine for lifting miners up shafts in stages. At first it consisted simply of large spikes driven into

a pump rod at intervals that corresponded to the height of the stroke. Small platforms were placed in the shaft so that men moving upwards stepped from rod to platform and awaited its return for the next upward stroke. The German engine was called a *fahrkunst* and was based on an earlier Swedish machine for hoisting ore. The *fahrkunst* man engine developed into a double-rod arrangement, with one rod rising as the other fell, and men stepped from one to the other as the rods paused between strokes. From the time of the first experiments in the Hartz Mountains, the Quaker Fox family, through the Royal Polytechnic Society of Cornwall, began to promote the introduction of something similar in Cornish mines. Mine owners and mine surgeons, like Dr Paul of Consols, had long struggled with the problem, and as early as 1829 Paul had encouraged Michael Loam, chief engineer at Consols, to solve it. Loam's experiments coincided with the introduction, in 1833, of the *fahrkunst* in the Hartz mines. Little progress was made in Cornwall during the 1830s, but the mines got deeper and hotter and mine owners and miners were suffering as a result. By 1840 miners were facing climbs of over 1500 ft (454 m) to and from their work places, and it was killing them. The Polytechnic Society offered a cash inducement to encourage mines to pursue a scheme to alleviate the situation, and in 1841 – undoubtedly partly prompted by the Foxes – Loam came up with his design. Tresavean mine responded and in January 1842 installed Loam's water-wheel-powered engine, which followed the *fahrkunst* double-rod design. The experimental engine carried men to and from the 24-fathom level without problem. Thereafter it was extended to the 248-fathom level, which was 1700 ft (515 m) from surface, and was then powered by a 36-inch cylinder steam engine. Three years later United Mines, another deep, hot mine, installed the same double-rod type of man engine, there powered by a 32-inch engine. For some reason no more were erected until 1851, when Fowey Consols installed a single-rod machine, which was powered by a 30-ft (9-m) water-wheel. Dolcoath, Levant and Wheal Vor all installed man engines in the 1850s, and during the 1860s ten more were installed between Devon Great Consols on the Tamar and Wheal Providence at St Ives. During the turbulent 1870s two were erected at Crenver & Abraham and South Caradon mine.[11]

The Cornish did not like the double-rod type of engine, believing it to be inherently dangerous, and so after the United engine was installed all engines were made to the single-rod design. Three man engines were water-wheel powered: the original at Tresavean and those at Fowey Consols and Cooks Kitchen, which used an existing 52-ft (16-m) diameter underground water-wheel for the purpose. This was later replaced by a 26-inch horizontal steam engine. All the engines except the one at Wheal Reeth (1861–69) were operated by cog-wheels, a crank and flat rods. Wheal Reeth's was unique in having the rod attached to the nose of the bob. The cylinder sizes of the steam engines varied between 19-inch and 36-inch diameter.

OPPOSITE
The Dolcoath Mine man engine. The engine for carrying miners to and from the surface was essentially a long rod from surface to near the bottom of the mine. As the rod moved up and down, the men stepped from fixed platforms in the shaft onto steps on the rod and off again when the rod stopped, thereby moving them swiftly to surface or to the men's working levels.

The Redruth and Chasewater Railway was opened in 1825 and operated until 1915. In 1855 two small steam locomotives replaced the horses that had been used to haul the wagons – they were named Miner and Spitfire and the one shown is Miner.

The longest haul was made by Dolcoath's engine, which, after several extensions and engine changes, brought men up from a depth of 351 fathoms or 2106 ft (638 m) at a speed of 83 ft per minute. It operated for over 40 years.[12]

In west Cornwall people still remember the man engine disaster of 1919 at Levant mine, in which 31 men were killed. Levant's engine served longer than any other in Cornwall, from 1856 to 1919. It was extended several times and the engine was replaced three times, the last having compound 18-inch and 30-inch cylinders. It finally collapsed in October 1919 after 63 years in service. Despite the accident – the second worst in Cornish mining history – the man engine was probably the biggest boon ever introduced in deep mines. It contributed directly to saving hundreds of lives and lengthened thousands more by cutting out the long, health-destroying climb to surface after an exhausting shift in strength-sapping heat.[13]

The first tramways and railways link mines to ports

An area in which Cornwall's industries lagged well behind other parts of the country was transport. Moving tin and copper ore across mines and to ports, and timber, coal, engine parts and other materials from those ports to the mines, was expensive and difficult. The long-established system of transporting such things by mule train was outdated and totally inadequate. The expansion of copper production in the first two decades of the 19th century necessitated a drastic improvement in transportation, and Cornish adventurers and engineers were well aware of the existence of tramways in other parts of the country. In the north of England tramways had been in use for over 100 years, and it is even possible that there were some track and flange-wheeled wagon haulage systems underground in Cornwall, although there is no record of their use at surface. As usual, Trevithick was one of the first to realise what was needed, though his inventiveness was well ahead of any practical application. By Christmas 1801 he had demonstrated a workable road locomotive and proceeded to display it on a track in various parts of the country. Shortly thereafter Lord de Dunstanville, who owned several of the mines where Trevithick was employed, promoted a plan to link his Camborne mines with his harbour at Portreath. The line was surveyed, but nothing came of it. In 1809 work began to construct a tramway from the North Downs mines to Portreath, and it was an immediate success. It was extended to Poldice mine in 1818 and was quickly earning a fortune from the rich copper mines of Gwennap and Redruth. An idea of the difference this tramway made can be gauged by the fact that before its inception it took something like 1000 mules a day to carry ore into Portreath, each animal carrying between two and three hundredweight. The tramway, with its three-ton horse-drawn wagons, was soon carrying most of the 25,000 tons of copper ore that went through Portreath annually.[14]

The success of the Poldice to Portreath line undoubtedly influenced John Taylor's plan for a similar but more ambitious scheme to link Redruth and Chacewater to quays at Devoran. This would benefit the whole of the district, including Taylor's enormous Consolidated–United group of mines, which were set to be worked on an unprecedented scale. An Act of Parliament of 1824 sanctioned the Redruth and Chasewater Railway, and in the following year work began to bring Taylor's plans to fruition. The new railway soon eclipsed the Poldice to Portreath line in the tonnage of freight it carried. During that time Taylor's Gwennap mines, employing over 3500 workers, became the largest industrial complex in the world. His new line carried a significant proportion of the world's copper ore and severely reduced the trade going through Portreath. Although the sea route from Devoran to Swansea was longer and ships had to go round the Lizard and Land's End, ports on the north coast such as Hayle and Portreath were often closed for long periods during winter by storms. The Redruth and Chasewater Railway was eventually nearly ten miles long after branch lines were added to serve Wheals Buller and Basset. In its first 51 years of operation the

Sir Francis Basset, Lord de Dunstanville of Tehidy (1757-1835). Basset was one of the wealthiest and most successful mine owners in Cornwall and when he died his funeral was attended by some 20,000 people from all walks of life and from all over the county. The miners appreciated his benevolence and fairness and contributed to the great monument to him that stands on Carn Brea, in the midst of his lands and mines. It was erected in 1836.

line carried 3,700,000 tons of freight, including 1,700,000 tons of copper ore and 1,500,000 tons of coal. During that half century it made a profit of £104,000, and in only one year was there a slight loss. The wagons were horse-drawn until 1855, when two small steam locomotives were introduced. Appropriately, they were called Miner and Smelter, and a couple of years later they were joined by Spitfire. The line finally closed in 1915, some 40 years after the demise of the great Gwennap copper empire it was built to serve.[15]

In 1837 work started on the Hayle Railway. This line was to run between Hayle and Portreath, with a terminus at West End Redruth and another at Tresavean mine. In later years its route was to be followed by the West Cornwall Railway and then by the Great Western Railway (GWR) and British Rail. The line linked the two most important ports on the north Cornish coast with what was again becoming the most important mining district. Hayle and Camborne were also the principal engine-building centres of Cornwall, containing Harvey's Foundry, the Cornish Copper Company, Holman Brothers and Tuckingmill Foundry. In addition to the great steam engines these firms built, they supplied tools and mining gear of every description to the mining districts of the world. The line ran from Hayle Foundry and Copperhouse to Angarrack, and then along the present railway to the western side of Barncoose, where it turned north across Illogan Downs to Portreath. It ended with a spectacular incline, complete with steam-powered cable haulage.[16]

OPPOSITE
The Portreath Decline. The Hayle Railway, started in 1837, ran from Copperhouse, Hayle to Portreath, with a branch to Tresaveam Mine at Lanner. The decline, seen here, was the most spectacular part of the system. A steam engine stood at the top of the decline and raised the trucks loaded with coal and gear for the mines and lowered them full of ore to be carried to the vessels in the harbour and thence to the smelters.

There are still extensive remnants of these historical mineral railways to be found, and industrial archaeologists see them, together with the scores of tramways on individual mines, as ideal routes by which to explore the unique mining landscape of Cornwall.

The genie of gunpowder – local manufacture and safety fuses

Since the late 17th century, when it was first introduced into Cornish mines, 'black powder' or gunpowder had been imported from other parts of the country and was expensive. Apart from the high cost of carriage from Somerset and other places of manufacture, the makers – conscious of their monopoly and Cornwall's need – added an extra premium. Some claimed, possibly without justification, that it was the cost of saltpetre (potassium nitrate, KNO_3) which made it so costly. By the beginning of the 19th century Cornish industrialists had plans to make gunpowder themselves, and in 1809 the first powder mill was established in Cosawes Valley at Ponsanooth. The location was perfect, for thick woods covered a narrow, steep-sided valley, which had an abundance of water to power the mills. The buildings were dispersed among the trees, isolating them from each other for safety. By the middle of the 19th century there were similar gunpowder factories in several mining districts, resulting in a dramatic reduction in gunpowder costs.[17]

A young Camborne miner lighting a safety fuze with his candle. The safety fuze was invented by William Bickford in 1831, and saved the sight of thousands and the lives of hundreds of miners in Cornwall.

From its first introduction into Cornwall the use of black powder had been accompanied by an unacceptably large number of accidents, many of which proved fatal. The burial registers of Breage record both the introduction of shothole blasting there by Thomas Epsley, in June 1689, and the first recorded death due to 'shuting the rocks'. John Archer was using gunpowder to blast underground at Trebollans mine in 1691 when he blew himself up and was killed. The expression 'shooting the rocks' was used in Cornwall until well into the 18th century. As the use of gunpowder spread accidental death due to its use became commonplace, and the loss of eyes or fingers and other injuries were almost daily occurrences. The methods used to introduce the flame to the powder charge were primitive, and throughout the 18th century the most usual way was to cut and join lengths of goose quill, which were then filled with finely crushed gunpowder. These 'fuzes' or 'rods', as they were known, could cause the main charge to explode prematurely, often resulting in death or injury.[18]

This undesirable situation changed dramatically with the invention of the safety 'fuze' (fuse) in 1831. William Bickford had moved from Liskeard to Tuckingmill, Camborne, where he ran a leather business. He was horrified by the many deaths and injuries suffered by miners due to blasting accidents, and as a devout Methodist he was determined to do something about it. He tried various designs consisting of parchment pouches containing a powder charge and fuse, but without success. Then one day he visited a fellow Methodist called Bray who operated a rope walk (rope-making works) at Tolvaddon. Watching Bray walk backwards spinning some yarn, Bickford had a brainwave. Why not pour fine gunpowder from a funnel into the centre of the rope as it was spun? Bickford, his son-in-law George Smith, who was a carpenter, Thomas Davey, a miner with an aptitude for solving practical problems, and Bray – all of whom attended the same Bible class at Tuckingmill – teamed up to produce a safety fuse that worked. This humane group of Methodists quickly perfected a design that not only worked easily and safely but was also simple to manufacture. The resulting safety fuse or 'safety rod' was patented in 1831 and after short-lived resistance became the standard blasting fuse in every mining field in the world. Tuckingmill, the village where it was invented, remained a centre of fuse manufacture until 1961, when ICI, the last owners, closed the factory and moved the operation to South Africa.[19]

As the 19th century progressed, both gunpowder mills and safety fuse factories were built in many mining districts in Cornwall. Large powder mills at Kennall Vale, Ponsanooth, and smaller operations like those at Cosawes, Herodsfoot and Trago Mills produced sufficient gunpowder for the needs of all of Cornwall's mines. Fuse factories under several names, such as Daveys at Gwennap, Bennets at Tuckingmill and Bruntons at Pool, competed with the Bickford–Smith factory to produce safety fuse. Until the introduction of electric fuses Bickford's invention remained the safest fuse to use, and it undoubtedly saved the lives of many miners in Cornwall and throughout the world.[20]

Four miners are hurrying away after lighting the fuzes on shot holes. Three fuzes can be seen spluttering as the miners move to safety. A fuze would burn at a foot a minute and the fuzes look to have been cut fairly short.

Mid-century: a mature, confident industry

The construction of extensive tramway and railway systems and the building of large foundries, gunpowder works and fuse factories were symptoms of an industry that was self-confident and optimistic about its future. By the middle of the 19th century the sheer scale of mining activity in Cornwall, facilitated by this support infrastructure, proved that the confidence and optimism were not misplaced. The statistics show that Cornish tin and copper production peaked in the 1860s. At that time more than 340 mines were operating in the county, 40 per cent of which worked exclusively for tin and 20 per cent for copper, with about 25 per cent producing a fair amount of both. More than a dozen mines produced lead, with wolfram, arsenic, blende, mundic, zinc and silver important by-products of mines in several districts. Iron ore was also mined in the central part of the county. Copper production peaked between 1855 and 1865, with an average annual output of 181,470 tons of ore. Tin metal was also at its peak of over 10,000 tons a year between 1863 and 1865, and although it was briefly to touch similar heights in the 1870s, thereafter the tonnage fell continuously. More than 40,000 workers were employed in these mines, with the majority (over 21,000) being employed in just 83 mines with between 100 and 800 employees. Although the three largest mines – Dolcoath, Devon Great Consols and Clifford Amalgamated – employed well over 1000 workers each, more than 40 per

A long-section of part of one of Dolcoath Mine's lodes. These working drawings, which show the east-west section and the transverse section of a lode, were used by mine managers to plan development and ensure adequate ventilation and economy of stoping.

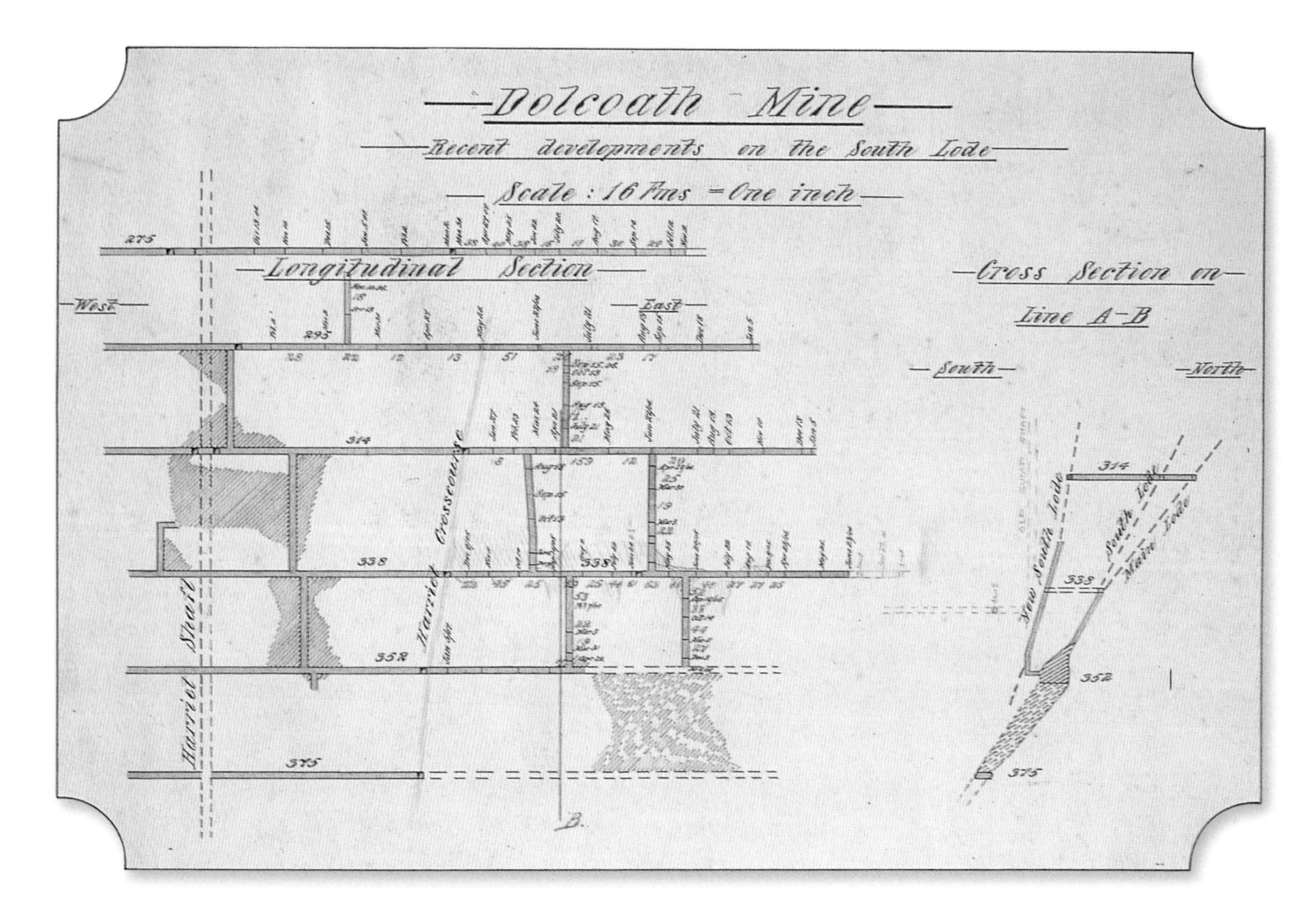

cent of Cornish mines employed fewer than 70 workers, and 25 per cent operated with less than 20.[21]

Sir Henry Thomas De La Beche (1796-1855). This engraving is by S. H. Maguire and is dated 1851. De La Beche wrote extensively on geology and was considered a leading authority on the subject in the 19th century. He was involved with the establishment of Geological Survey of Great Britain.

The machinery on these mines – most of it manufactured in Cornish foundries – was truly impressive. More than 600 steam engines were working in the 1860s. These varied in size from small whim and capstan engines with a cylinder diameter of perhaps only 10 or 12 inches, to great pumping engines with cylinder diameters as large as 100 inches. These engines not only pumped vast volumes of water from as deep as 2000 ft (606 m), they hoisted ore and operated man engines, crushers and stamps. The Cornish beam engine had been brought to perfection by the pupils of Trevithick and Woolf, and engineers such as William and James Sims, William West and Samuel Grose were designing and installing engines that were to work efficiently and economically for decades, in some cases well into the 20th century. More than 220 mines used at least one steam engine, 44 had steam- and water-powered engines, and many mines relied totally on water power. In addition to all this machinery, almost all the tin mines had water-driven stamping mills. Some mines employed an enormous number of engines of all types: Fowey Consols had six steam engines, 17 water-wheel engines and three hydraulic-pressure engines. Clifford Amalgamated had 11 steam pumping engines and numerous others for hoisting and crushing. Par Consols had 15 steam engines and Devon Great Consols operated seven steam engines and 32 water-wheel engines. Some small mines merely used tiny and primitive 'flopjack' pumps to drain their workings, while others worked no deeper than adit level.[22]

As the above facts and figures show, it is impossible to generalise and portray a typical Cornish mine of the mid-19th century. Just as the numbers employed and machinery used varied from mine to mine, so did the mines themselves. In 1865 half-a-dozen mines had gone deeper than 1800 ft (545 m), with Fowey Consols bottoming out at 2040 ft (618 m) and Dolcoath at 1920 ft (581 m). Although most mines operating in the 1860s had reached 1000 ft (300 m), at least 80 mines went no deeper than 300 ft (90 m), and many worked at or just below adit drainage level. Were we to include the scores of mines that had closed during the previous 20 years or so – many of which opened and closed several times during the 19th century – we might find that a majority of Cornish mines never went deeper than 500 ft (150 m). The same variety is noted with respect to the extent of the workings. Carn Brea, Dolcoath and Clifford Amalgamated stretched for a mile or more and worked several different lode structures, whereas dozens worked a single lode for a hundred feet or less. Surviving plans show that small mines often had a couple of levels with

BAL MAIDENS

'Bal maiden' or 'Bal maid' are generic terms for women or girls who were employed at the surface on Cornish mines. Their tasks were almost always to do with reducing, separating and dressing the ore. In copper mines these processes were carried out before the ore was sampled and assayed; in tin mines they followed stamping and involved washing, or 'buddling', away the waste sand and slime to produce the 'black tin' concentrate. The earliest references to female labour on the surface at mines in Cornwall and Devon come to us from the end of the 13th century, when the Bere Alston and Bere Ferrers mines on the Tamar River were established to mine silver for Edward I. The mine wage accounts for May 1306 contain the names of perhaps the earliest identified Cornish 'bal maids': Agnes Oppehulle, Isabelle Cutard, Gunhild daughter of Bon, Dionis' de Milleton and Emma de Fallyng. They were all employed 'washing black ore' at Calstock. The extensive stannary records for the medieval period contain hardly a hint of female involvement in tin dressing, although it would be surprising if they did not work occasionally with their menfolk at such tasks when the need arose.[25]

A group of Camborne ore dressers including seven bal maidens in about 1900. The ladies are all wearing the traditional 'gook' headgear.

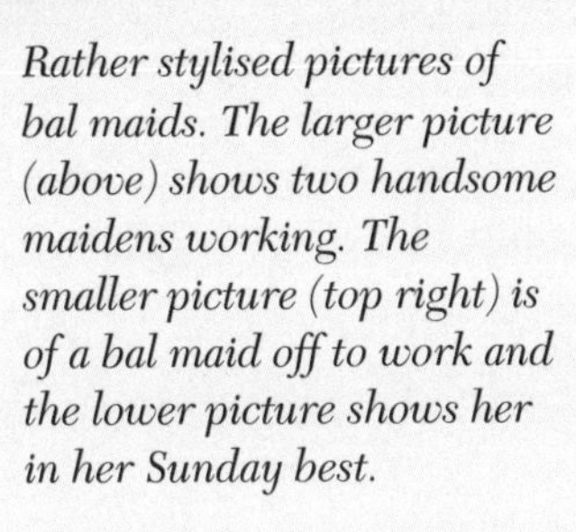

Rather stylised pictures of bal maids. The larger picture (above) shows two handsome maidens working. The smaller picture (top right) is of a bal maid off to work and the lower picture shows her in her Sunday best.

a total distance driven of just hundreds of feet. In contrast, great mines measured levels in miles: Gwennap's United and Consolidated mines had over 100 miles of tunnels. There was also considerable difference in the size and value of lodes worked.[23]

Female labour on the mines

The period when women and girls were to play an important part in surface work at mines was during the years of the copper boom between the beginning of the 18th century and the second half of the 19th century. The earliest extant, detailed copper mine accounts are from the 1720s and 1730s, and these show that large numbers of bal maids were at work on the copper floors of Wheal Dudnance, Penhellick and Trevenson. Attempts to establish the numbers of these female workers have proved fruitless, for there are no reliable figures from the 18th century with which to work. Thus, John Rowe's estimate of 1200 to 1800 for the year 1787, John Vivian's figure of 4000–5000 for 1799, Spackham's figure of 2276 for 1827 and Sir Charles Lemon's of 4414 for 1836 are at best educated guesses and really only hint at the number of bal maids on Cornish mines. Census returns from the mining parishes, Parliamentary Commissioners' reports and other sources of material for the decades between 1840 and 1890 are equally vague and can give figures that vary from as low as 4135 in 1841 (census) to as high as 14,400 in 1843 (Watson). What is certain is that very large numbers of women and girls worked alongside large numbers of men and boys to bring the copper ore to a merchantable condition. In the early 1860s more than 2500 bal maidens were working within five miles of Camborne. At the peak of copper mining between 15 and 20 per cent of all mine workers were female. In 19th-century tin mines the proportion might have been as low as 2 per cent.[24]

Five basic tasks were carried out on the dressing 'floors' of copper mines: 'ragging', 'riddling', 'spalling', 'cobbing' and 'bucking', the last four of which fell to the bal maids. Ragging was carried out by youths using 14-lb (6.3 kg) sledgehammers to break down the larger rocks. Next, girls of about 16 years would use a sieve-like apparatus to riddle or griddle the broken ore. The separated rocks were then beaten to fist-size by girls of a similar age using a spalling hammer weighing between 5 and 7 lb (2.3 and 3.2 kg). In the next stage girls of about 15 used a cobbing hammer to break away the ore from unwanted gangue material. This hammer had an unusual shape, having a long head which curved forwards rather than back, as with a pick. Finally, the ore-bearing rock was subjected to bucking. The best-paid job, this was reserved for more experienced women, who could earn up to a shilling a day in the 1850s. A square-headed, flat-faced hammer was used by the bucker, who placed the gravel-sized rock on a metal plate or anvil. The bucker held her hammer with both hands, striking and grinding the gravel in one fairly deft movement. Other tasks of bal maidens were barrowing rocks across the dressing floors and general labouring.[26]

Descriptions of Cornish bal maidens by 19th-century essay writers and other visitors to Cornish mines have done their reputation no favour. Well-meaning middle-class do-gooders with liberal ideas on working-class behaviour have painted a distorted picture of these proud, independent, self-confident and hard-working women. They were undoubtedly rough and ready, they were often outrageous in their behaviour – especially (as contemporary journals show) when copper agents were on the mine sampling – and some were foul-mouthed and loose in their morals. Maybe so, but such generalisations could be applied to almost any group during the Industrial Revolution. Many were also devout church- and chapel-goers. The bal maidens' reputation for spending their hard-earned surplus cash on trifles, cheap clothes and visits to the fair is justified. It was their money, they had earned it and they spent it how they pleased. What is most important about these women is that they chose to be bal maidens even when better-paid and easier jobs were available. They were not, as some breast-beating modern historians insist, victims. They were not exploited. They chose a hard life on the mine – just as their menfolk underground chose theirs – because it was the kind of life they wanted. They enjoyed the singing, the banter, the camaraderie and the challenge. These tough, courageous, free-thinking women would have been insulted by those patronising historians whose view of the past owes more to the 'heart on their sleeve' than to the reality of how and why our ancestors did what they did.[27]

The threat from abroad

By the 1860s what had been a merely worrying straw in the wind was becoming a hard, threatening reality. Copper and tin were being discovered and exploited across America and the colonial territories. The Cornish copper bubble was about to burst. Ironically, it was mainly Cornishmen who were discovering these new sources of mineral wealth, and it was their 'Cousin Jacks' who were flocking to exploit them. (This nickname was acquired in America because of expatriate Cornish miners' supposed habit of asking for a job for 'Cousin Jack' back home in Cornwall.) As large deposits of shallow, metallic ores were found and and could be worked by cheap local labour, capital raised in London was being redirected abroad to earn larger, quicker and safer returns for investors. As a result of these vast new orefields the prices of metals fell inexorably, and the hitherto booming Cornish copper mining industry quickly found itself in crisis.[28]

CHAPTER EIGHT

CHANGE AND CRISIS

Late 19th century (1866–96)

OPPOSITE
Machinemen operating a heavy drifter machine at Dolcoath Mine in 1900. Note the miner spraying the face with water to lay the dust and the candle stuck by clay to the rock. These drifters became known as 'bar-and-arm' machines, as they were attached to the vertical steel bar by a horizontal steel arm.

FOLLOWING THE COLLAPSE of the copper price in the 1860s and the consequent crisis in the traditional copper parishes of west Cornwall, the picture became confused as areas to the east of Truro continued to be productive and appeared to buck the trend and Devon Great Consols went from strength to strength. From an average copper metal price of £136 per ton between 1855 and 1860 the price dropped to an average of £105 per ton in the years 1865 to 1870 and to £95 between 1875 and 1880, the lowest for 90 years. By the 1890s it had dropped to an incredible £38 a ton. The reasons were not hard to find: Chile produced nearly 90,000 tons of copper ore in the 1840s and 250,000 tons in the 1850s. By the 1860s Bolivia and Peru were contributing to South American production and the tonnage doubled again.[1]

Mixed fortunes in the second half of the 19th century

In the 1850s, as the effects of this fast-growing New World output were beginning to make themselves felt in the Camborne–Redruth–Gwennap area, fresh discoveries had seemed to bring new hope to the area. It was, unfortunately, only a temporary reprieve. During the 1840s, as the grade of copper ore fell and deep mines struggled, shallow deposits of good copper ore were found at Wheal Buller, where a small outlay of £1280 produced dividends of nearly £100,000 in the years 1853 and 1854, and by 1861 the mine had made a clear profit of £244,000. On the same lodes to the west of Wheal Buller, Wheal Basset also made a huge profit during those years, paying out to her adventurers nearly £300,000. Further still to the west lay South Wheal Frances, which between 1846 and 1860 made a profit of £180,000 from copper sales. North and West Basset mines also produced large tonnages of good-grade copper ore in the 1850s. These mines lay in the vale between Carn Brea and Carnkie – for centuries rich in tin ore, but until that time hardly touched for copper. By the end of the 1860s, however, all of them were once again turning to tin as the potential of the Great Flat Lode became apparent, and the whole range of mines spread along the south flanks of Carn Brea, Carn Arthen and Carn Entral between Redruth and Troon began to open up one of the most productive tin lodes in Cornwall. Over the next half century or so, from Wheal Uny in Redruth, through the great Basset mines at Carnkie, to Wheal Grenville at Troon, the Great Flat Lode

produced millions of pounds worth of tin ore and the mines along it employed many hundreds of workers.[2]

The copper mines to the east of Truro were also badly affected by declining grades, expensive deep workings and the falling copper price. At Fowey Consols production fell from 5898 tons of copper ore in 1855 to a mere 480 tons in 1867, and she never really produced much tin. Par Consols was virtually finished by 1869, and although she mined some tin ore it was insufficient to keep the mine going into the 1870s. Further east the picture looked more promising, with the Caradon mines continuing to produce copper ore into the 1880s. Caradon Consols sold 2966 tons of ore in 1873, but declined thereafter until it closed in 1885. East Caradon peaked with 5098 tons of copper ore in 1865, but by 1885 this had fallen to a mere 52 tons and the mine was shut down. South Caradon mined 6468 tons of copper ore in 1877, but there too output declined to just 162 tons in 1886, when it closed. West Caradon went down from 4457 tons in 1845 to 20 tons of copper ore in 1886, when it too finally closed.[3]

In the far west, in St Just parish, copper mining continued to be significant until the end of the century. Levant produced copper ore throughout the period 1820 to 1900, and during the last two decades of the 19th century her output actually rose,

Wheal Cock section of Botallack Mine, St Just. This derelict scene became typical all over west Cornwall as the mines closed one after the other throughout the early decades of the 20th century.

Devon Great Consols, made up of Wheals Maria, Fanny, Anna-Maria, Josiah and Emma, was the greatest copper mine in southwest England duing the second half of the 19th century.

until by 1900 it stood at 5064 tons. From the middle of the century Levant was also a significant tin producer, selling 665 tons of black tin in 1893 and 464 tons in 1900. Botallack mine, next to Levant on the cliffs, also produced copper ore until the 1890s, but production was in constant decline after the 1850s and she was principally a tin mine for most of the second half of the 19th century. By the end of the century the workings of these ancient mines stretched for half a mile beyond the cliffs. Both St Just mines were also significant producers of arsenic at the end of the century.[4]

The giant next door – Devon Great Consols

The mine which almost single-handedly kept the flag flying for copper in the south-west during the second half of the 19th century was Devon Great Consols, on the Devon side of the River Tamar. Between 1844 and 1903, when it closed, the mine produced nearly 750,000 tons of copper ore. Although this figure dwarfs the 19th-century output of even the great Dolcoath, Consolidated and United mines, it must be appreciated that all three of these mines had produced vast tonnages long before accurate records were kept. This fact notwithstanding, in just one year (1855–56) Devon Great Consols' output was greater than the total annual tonnage of Cornish mines in the 1790s. Devon Great Consols consisted of the original sett of Wheal Maria, together with Wheals Fanny, Anna-Maria, Joseph and Emma. These mines stretched for two miles along the largest sulphide lode ever discovered in the southwest. The lode varied in width from 20 to 40 ft (6 to 12 m), and as it was worked away it left vast, empty gunnises, which required enormous timbers to support the wall rock. The lode had long been known about by local miners, who, led by Captain Richard Clemo, tried in vain to obtain the lease to work it. In 1844, however, Captain Josiah Hitchens gained enough support to form a cost-book company (see pages 154-55) and took the lease. The company was formed with 1024 £1 shares, and, with no further 'calls' until the company's reconstruction in 1872, £2,500,000 worth of copper ore had been sold and £1,000,000 distributed to the shareholders. Within a few years of starting up the mine boasted a greater array of machinery than any other copper or tin mine in the southwest, with 33 water engines, eight steam engines and two man engines. It employed 1300 workers, with no less than 450 men and boys working underground supervised by 20 mine captains. On the copper floors worked 217 females, 168 boys and 136 men. Enginemen, mechanics, carpenters, coopers, blacksmiths and office workers made up another 300 employees. Devon Great Consols had its own foundry and smithshop, as well cooperage, sawmills and stables for its many horses. It even ran its own school for the children of its employees. Initially horses and carts were used to convey ore to the Tamar River, but later a five-mile long railway was constructed between the mine and the quays at Morwellham. This eventually employed three locomotives hauling 60 wagons to carry ore to the dock and coal, timber and other gear back to the mine. By the time it closed in 1903 the mine had raised and sold ore to the value of over £4,000,000 and paid out £1,250,000 to its adventurers.[5]

Tin makes a comeback

Meanwhile, the changes taking place in the Central Mining District (Camborne–Redruth) were profound. By the middle of the century, with the quality and the quantity of copper ore dropping and tin being discovered in ever larger quantities at depth, the parishes which had been at the centre of world copper production were transformed into an important tin-mining area. Although tin dominated in the second half of the century, we have to return to the beginning of the 19th century to discover the root causes of the many changes in the industry. The Association of Proprietors of Tin, set up in 1780 to regulate the price of the metal, had some limited success, but by the 1830s it was generally failing for a variety of external reasons, and the abolition of coinage in 1838 was to change the industry for ever. The increasing demand for tin by the Welsh tin-plating industry, which had been growing since the late 18th century, led to this industry replacing the East India Company as the prime customer for Cornish producers. Nine works operated in 1800, and by 1825 this had grown to 16, with output of tin plate quadrupling between 1805 and 1837. These Welsh customers quickly became all-important to the Cor-

A CAMBORNE MINING DYNASTY

Throughout the 18th century the mines of Camborne were dominated by just a few families who largely owned and organised them. They were the John, Vivian and Trevithick families. By the end of the century, the Johns had mostly moved on and the Trevithicks were involved more in engineering than in mining, but the Vivians were still there, along with their relatives, in-laws and friends. However, 19th-century Camborne was to be dominated by a new family of mining men – the Thomases.

When Dolcoath reopened in 1799, Captain Andrew Vivian and his relative Richard Trevithick were among those who persuaded Lord de Dunstanville to grant the lease of the mine. Vivian was manager and Trevithick chief engineer. But within a few years Andrew Vivian had handed over the management to another relative, Captain Rule, and Richard Trevithick was already loosening his ties with Camborne. Rule was there as manager between 1806 and 1834, when he handed over to his nephew, William Petherick. Petherick's great friend and mentor was Captain James Thomas, brother of Captain Charles Thomas, and when Petherick died suddenly at the age of 47 in 1844, the nephew of James and son of Charles took over the management of the mine.

This man, Captain Charles Thomas Junior, was the most famous mine manager of his day. He is credited with being the first 'modern' mine manager in Cornwall due to his systematic approach to planning, working and organising Dolcoath. His great-grandfather had moved to the Camborne area as a tin streamer and miner, his grandfather had been a miner at Dolcoath and his father and uncle were mine captains there, so it was only natural that he should follow in their footsteps. By the time of his death in 1868 he had supervised the complete transformation of Dolcoath – which had been a failing copper mine when he became manager – into what was the most successful and profitable tin mine in Cornish history.

When he died his son Captain Josiah Thomas was appointed manager, and he in turn became the most famous and highly respected mine manager of his age. He travelled the world to advise on mines and mineral deposits and wrote innumerable papers on mining practice. Like his father, uncle and grandfather he was a devout Wesleyan and respected local preacher. When Josiah died in 1901 the miners wanted to erect a statue of him, but they settled instead for a memorial hall, which still stands in the centre of Camborne.

Josiah was succeeded as manager of the Dolcoath mine by his son, Arthur Thomas, who managed the mine through the many crises of the early decades of the 20th century before finally giving up mine management in the late 1920s. Arthur's son, Leonard, replaced him as Chairman of the Board at the Camborne School of Mines and for a while was managing director of South Crofty mine, which included the old Dolcoath mine workings. No mining family in Cornwall had played a greater part or had such high reputations in the industry as did the Thomases of Camborne. From the middle of the 19th century they dominated Cornish tin mining, just as the Lemon, Williams and Vivian families had the copper mining industry during the previous century and a half.[16]

nish tin mines, and they were to remain so for the rest of the century. Unimaginably large deposits of alluvial tin in the East Indies were already profoundly affecting the tin price, and as early as 1813 the East India Company transported 1700 Chinese labourers from Canton to the island of Banca to work the streams there. By 1816 tin exports from these deposits into China were 200 tons, with that from Cornwall totalling 350 tons. Worse was to follow, as for the first time Straits tin was being imported into England by the East India Company, which made a handsome profit from it. Before long Chinese merchants had taken over the supply of tin from the East Indies to China and the Cornish were forced to reassess their markets and look elsewhere for buyers. Thus, the Welsh tin-plate industry, rather than being just a welcome bonus to the Cornish tin-mine owners, became their lifeline. The violently fluctuating price of tin metal during the first half of the 19th century reflected several dramatic changes in the industry. In 1800 the price was £101 a ton, and there was a gradual rise to £120 by 1806 and a rapid increase to £157 in 1810. The price remained high until 1815, when it collapsed, falling to £73 a ton by 1820. Thereafter the price fluctuated between £95 and £73 until the middle of the century.[6]

Undoubtedly, the tin mine that dominated the first half of the 19th century was Wheal Vor in the parish of Breage. In 1823 it was reported that Wheal Vor 'recently made in three months a profit of £13,000, the value of its tin alone in that time amounted to £25,000 – so rich a load of that metal has not, we are informed, been ever before worked in Cornwall – its width is said to be thirty-six feet'. Wheal Vor was an amalgamation of four mines all working the same lode – Wheal Vreah, Carleen, Polladrass and Wheal Vor. In 1823 the mine employed some 541 men and boys underground, besides a considerable workforce on the surface. So large was its output that in 1820 Wheal Vor in effect had its own smelters, using the old copper furnaces of Sandys, Carne & Vivian at Copperhouse, Hayle. It was not long before this great mine moved its smelting operation on to the mine itself, and throughout the 1820s it produced the largest tonnage of tin metal of any Cornish smelting house: between 1824 and 1830 Wheal Vor mined and smelted a quarter of all tin produced in Cornwall. Legal disputes and other difficulties caused problems for the mine and by the middle of the century it was struggling. During the 1850s Wheal Vor was not on the dividend list of profitable Cornish tin mines, but by 1861 it had recovered sufficiently to be in profit, remaining so for several years, and even replacing the revived Dolcoath as the most profitable tin mine in the years 1864 and 1865.[7]

Teaching the world to mine – the Cornish diaspora

Already by the start of the 19th century groups of Cornish miners had migrated to mining fields across the British Isles and abroad. By the end of the century there was not a mining field in the world where Cornish engineers and miners were not prominent. The early decades of the 19th century saw Cornish miners working the copper mines of western Ireland and South Wales. The Vivians and Grenfells, Cornish industrial giants, were already established in Wales, as the Williams were in Ireland. The export of high-pressure Cornish steam engines to the silver mines of Peru, overseen by Richard Trevithick between 1814 to 1818, was the start of the transatlantic migration of mining technology and skilled miners to former colonial regions, bringing them into the modern, integrated, global mining economy. This world of mining, with its interdependent finance, labour and technology, was largely being shaped by the efforts of Cornishmen.[8]

John Samuel Enys (1837-1912). The Enys family were great mineral owners in Cornwall, and several members of the family were prominent mine adventurers.

In the gold and silver mines of Latin America Cornish miners not only taught the natives their skills but learned new techniques themselves as they mined unfamiliar ores under conditions they had not previously encountered. The reputation of the Cornish as being able to work any type of ground under the most arduous conditions soon established them as the finest hard-rock miners in the world. No ground was too hard, too unstable, too wet or too soft for these men to mine it successfully. They truly 'taught the world to mine'. They led the way in mining lead in Wisconsin and Illinois, gold in the Appalachians, lead in Spain and the Isle of Man, copper in Ireland, Norway and the Caribbean – in fact almost any and every metallic ore discovered throughout the globe. Without Cornish knowhow and machinery the Californian gold rush would have stalled a short way below surface and been delayed for half a century. During the 1840s and 1850s, as the mineral wealth of the USA was opened up, it was Cornishmen who discovered the ore, opened the mines, and managed and worked them as skilled miners. Zinc in New Jersey, gold in Montana, Colorado and the Dakotas, copper in Arizona, Montana and Michigan, silver in Nevada and Utah and gold and mercury in California were all exploited by these pioneers of American mining. They were among the first to head into the frozen wastes of northern Canada, many of them dying in tents a thousand miles from civilisation. No mining task was too difficult, too isolated or too dangerous for them.[9]

As in America so in the Antipodes, the Cornish took the lead in discovering, opening up and working the mines, whether they were for copper, gold, silver, zinc or tin. South Australia, Victoria, New South Wales, Tasmania and Western Australia were joined by New Zealand and the South Pacific on the itinerary of the Cornish miners as they worked their profitable way around the globe. Former boom towns such as Bendigo, Ballarat, Kadina, Kapunda, Moonta and Bura Bura exist today because of the work of these 19th century-Cornishmen. As in Wisconsin, there are more Cornish names in South Australia than there are in parts of Cornwall.[10]

By the last decades of the 19th century there was a well-worn route to South Africa as the gold and diamond mines there were opened up. Copper in Namaqualand in the 1850s, diamonds in Kimberly in the 1870s and gold in the Transvaal in the 1880s were all exploited by the ubiquitous Cornish miner, who appeared as if by magic at the first indication of mineable ore. Before the Boer War one in four of all white miners in southern Africa was Cornish. During the conflict many returned home, but many others joined the volunteer battalions that fought the Boers. By the outbreak of the war Cornish miners had already moved into Central Africa, East Africa and West Africa to work for gold, tin and copper, before spreading out to India and South East Asia in their insatiable desire to seek out the riches that lay underground. Many of the mines they established and the engines they installed remain as reminders of that tremendous period of activity during which a tiny rural population from a small corner of Britain helped to change the face and character of whole regions of the world.[11]

OPPOSITE
Painting by Terence Cuneo of cylinder head of Robinson's 80-inch engine at South Crofty Mine. Designed by Samuel Grose and built in 1854 by Sandys, Vivian & Co. for Alfred Consols Mine at Hayle, the engine moved around several mines before being purchased by South Crofty in 1903. She worked almost without interruption until being replaced by electric pumps in 1955, by which time she was pumping water from over 2000 ft (610 m) below the surface.

In Great Britain itself Cornishmen helped in the development of the Cumberland and Lancashire iron mines from the 1850s, having already vastly improved lead-mining techniques in Wales and Derbyshire. John Taylor took Cornish mine captains and skilled miners to several Pennine mining fields to facilitate the opening of formerly unproductive lead mines, which they did successfully. Others moved north to work in coal mines and ancillary industries.[12]

Strides in mining and processing technology

The second half of the 19th century saw many changes in technology and mining practice. Improved steam-powered machinery was being harnessed to every conceivable process as larger tonnages of ore were being hoisted up well-constructed vertical shafts using wire ropes and skips of increased capacity. Cages for man-riding, sometimes equipped with fail-safe devices, were becoming normal on the larger mines. Surface plant for mineral processing was constantly improved and became more sophisticated as revolving Californian stamps and air-cushion stamps were introduced along with Frue vanners and shaking tables for ore concentration. All of these improvements – some introduced into Cornwall by Cornishmen working abroad – helped make Cornish mines more competitive.[13]

Underground too there were many changes in machinery and practice. Since the introduction of gunpowder for blasting in the 17th century all shotholes had been hand-drilled. Three men in an underhand stope could expect to drill two or three 4-ft (1.2 m) holes in a shift. From the 1860s this situation began to change after Frederick Doering introduced a compressed-air rock-boring machine to the Cornish mining fraternity. After witnessing a successful demonstration at the Polytechnic Society's exhibition of 1867, Captain William Teague, the go-ahead manager of Tin-

croft mine, asked Doering to demonstrate his machine at Tincroft. Teague was impressed with the results, which were twice as efficient as drilling by hand. In December 1869 Doering began a similar demonstration at Dolcoath. After 18 months his machine had had only limited success, but it was not long before other designer-engineers tried their hand and another machine, the Barrow rock drill, proved more successful. The Barrow was patented by Cornishmen working in Barrow-in-Furness, Lancashire, and was tried at Dolcoath in 1878. It proved to be efficient and advanced a level far faster than could be done with hand labour. It was made of gunmetal, was reciprocating and weighed 120 lb (54 kg). In operation it was mounted on a 'bar-and-arm' (a metal bar fixed vertically between floor and roof with an extended arm to which the machine was attached) and could be manoeuvred easily to drill holes at various angles. It worked efficiently with compressed air to 50 psi. It was not long before the Barrow rock drill was in use in other Cornish tin mines, where it proved a boon. Quickly other Cornish firms – notably R. Stephens & Son (makers of Climax rock drills) and Holman Brothers – became involved in modifying and patenting designs, as did Tuckingmill Foundry and Sara's Foundry at Redruth. Some worked well and others fell by the wayside, but the last decades of the 19th century saw an explosion of inventiveness in the field of compressed-air machinery, in which Cornish firms like Holman and Stephens took the lead. These Cornish companies made a variety of drilling machines, including heavy 300-lb (136-kg) 'drifter' machines; many went for export to mining fields all over the world.[14]

Coincidental with the introduction of compressed-air rock drills was the invention of dynamite. From the 1870s this high-explosive had proved, effective, efficient and relatively safe. The main problem was its high cost in Cornwall. In Germany and Sweden, for example, it cost far less. So most mines contributed to a fund to fight the 'dynamite patent', which was thought to be responsible for the high price. Eventually, as with gunpowder nearly a century earlier, factories were set up in Cornwall to manufacture dynamite and the price paid by the tin mines fell. The most famous factory was at 'Dynamite Corner', Upton Towans, where the National Explosive Works was built. Others were at Trago Mills and Herodsfoot – both formerly gunpowder factories – and at Cligga Head, Perranporth. The Bickford–Smith, Bennets and Bruntons factories continued to make safety fuse, but for use with dynamite this was equipped with detonators.[15]

OPPOSITE
Miners stoping at East Pool Mine, Illogan, in 1893. This scene was typical of how tributers worked over the centuries, using hand labour to 'beat the borer'. It was normal for one man to hold the drill steel while two men beat it, using spalling hammers of 5 to 7 lb (2.3 to 3.2 kg) weight. Note how the stope was lit by candles placed around the walls.

A great Cornish mine – Dolcoath

Just as Devon Great Consols dominated copper production in the southwest during the second half of the 19th century, so Dolcoath completely dwarfed all other tin mines in Cornwall. After several years of struggling with low copper production and only slowly increasing tin output, Dolcoath emerged in 1853 as a serious, long-term producer of tin. In that year the great mine paid its first tin dividend, and by 1859 it had leap-frogged all others and taken a commanding lead in tin production. Within a couple of decades Dolcoath's depth, wealth, profits and skilled workforce had become legendary, and the mine was known throughout the mining world for being innovative, advanced and successful. The expression 'deep as Dolcoath', and others slightly less polite, reflect the way this great mine was viewed in Cornwall and the mining world in general. By the end of the 1890s Dolcoath had reached 3000 ft (909 m) below surface and was being approached by means of a new vertical shaft – Williams Shaft – on the northern flanks of Carn Entral.

The extent of her importance can be gauged by the production figures: in the five years between 1884 and 1888 Dolcoath produced 12,421 tons of black tin concentrate. Her nearest rival, East Pool, sold 6937 tons in the same period. Between 1853, when she divided her first profits from tin, until 1895, when she was reconstructed as a limited company, Dolcoath raised and sold just under 59,000 tons of black tin. During the final 25 years of the old cost-book company the mine paid out £580,000 in dividends. In the last quarter of the 19th century the workforce rarely fell below 1200 and by 1900 it stood at 1379 workers, 773 of them underground. Her management, in the capable hands of the Thomas family, was respected wher-

Miners at East Pool Mine, Illogan, drilling an 'upper' in the back of the level in 1893. Note the candle on the end of the hammer handle, shining on the top of the drill steel. Cornish miners prided themselves on being able to drill holes at any angle in any rock.

ever metalliferous mines operated, and no miners or mining engineers were more worthy of the saying that 'Cornishmen taught the world to mine' than the men of Dolcoath. The name Dolcoath was as familiar in California as it was in Moonta or the Rand. Following its appearance on the scene as a profitable tin mine in the 1850s, Dolcoath did not lose its position of dominance until the Great War cut it down, as it did so much of British industry. Wheal Vor, Carn Brea, East Pool, Tincroft and Polberro all produced large tonnages of tin, as did a host of smaller mines, but Dolcoath, along with its neighbours east of the Great Crosscourse, carried the flag for the Cornish tin industry for the whole of the 70 years before the industry's near collapse in 1920.[17]

Foreign tin plays havoc

The progress and success of tin mining in Cornwall during the second half of the 19th century, however, was not without its problems. In the 1860s, when tin production was surging ahead in the county, anarchy and piracy plunged Malaya and the Dutch East Indies into chaos, and by the early 1870s production in Malaya's vast alluvial fields was at a standstill. The price of tin went through the roof, reaching £153 a ton, and in all the stanniferous districts of Cornwall tin mines were opened, reopened or expanded. Then, just as dramatically, it all went wrong again when the Royal Navy intervened in Malaya to restore order and allow the tin industry there to resume operations. In 1874 the price of black tin collapsed to £56 a ton, and most of the newly opened Cornish mines closed. Four years later tin was selling at £35 a ton, making most of the rest uneconomical and forcing even the richest to make economies and tighten belts. For a while Dolcoath itself looked shaky. At about the same time tin was discovered in Australia and Tasmania – ironically by Cornishmen – and Cornwall's problems multiplied as skilled men rushed out to exploit the new discoveries. The twenty years leading up to 1896 were among the most difficult in the long history of Cornish mining, and by the end of that period, after severe fluctuations in the tin price, the figure settled down at £64 a ton. Only the biggest and richest mines could survive under such circumstances.[18]

Labour problems

The final decades of the 19th century were not without problems of an entirely different kind – some arising from long-felt resentments, others originating externally. The five-week month – the traditional division of the year by cost-book companies – resulted, according to the miners, in a loss of four weeks' pay every year. For many 'owner's account' men and some tutworkers the claim was undoubtedly justified. By the 1870s this was no longer acceptable to many miners, especially those back from well-paid jobs in America. The manipulation of tutworkers' contract rates had long

ADVANTAGES AND DISADVANTAGES OF THE COST-BOOK SYSTEM

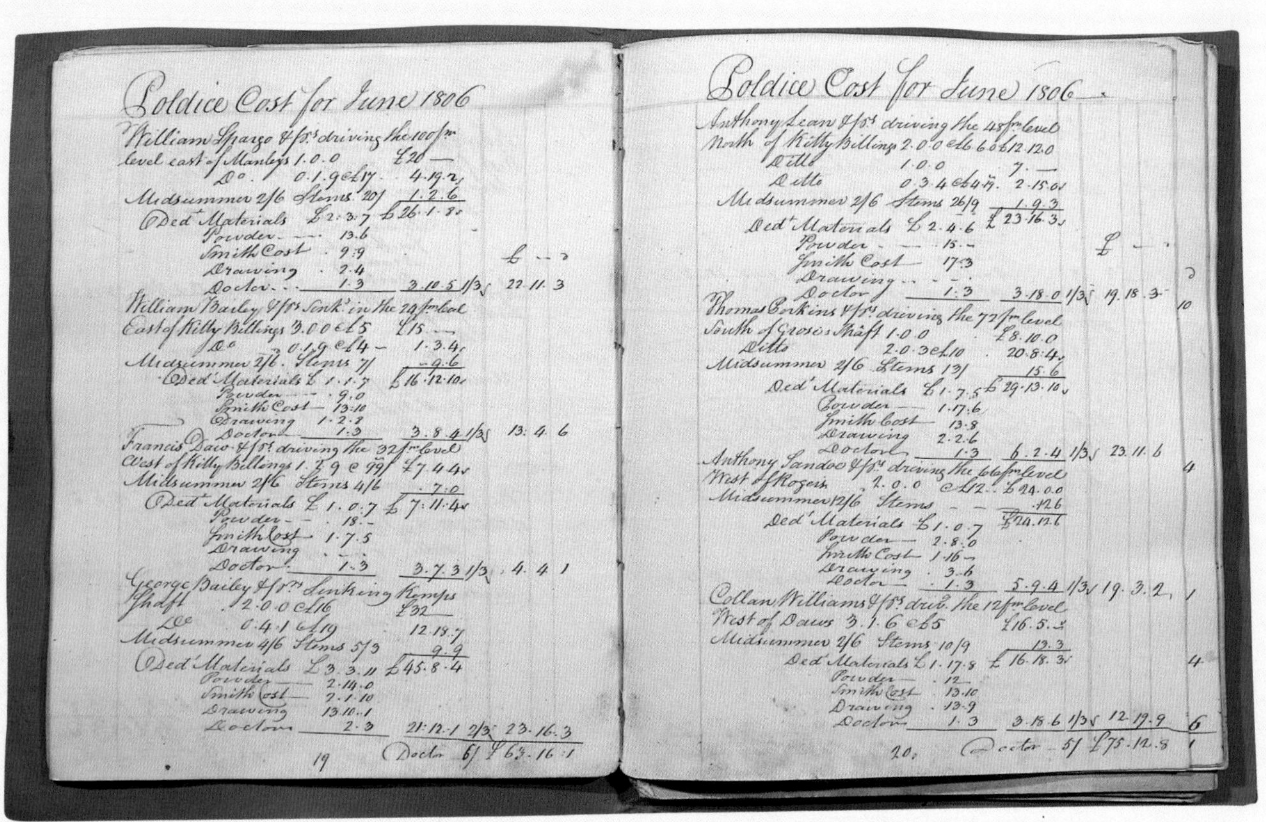

Poldice Cost for June 1806

William Spargo & Prs driving the 100 fm level east of Manleys 1.0.0 £20 —
Do. 0.1.9 @ £17 4.19.2
Midsummer 2/6 Stems 20/ 1.2.6
£26.1.8
Ded. Materials £2.3.7
Powder 13.6
Smith Cost 9.9
Drawing 2.4
Doctor 1.3 3.10.5 1/3 22.11.3

William Bailey & Prs Sink. in the 24 fm level East of Kitty Billings 3.0.0 @ £5 £15 —
Do 0.1.9 @ £4 1.3.4
Midsummer 2/6 Stems 7/ 9.6
£16.12.10
Ded. Materials £1.1.7
Powder 9.0
Smith Cost 13.10
Drawing 1.2.8
Doctor 1.3 3.8.4 1/3 13.4.6

Francis Daw & Prs driving the 32 fm level West of Kitty Billings 1.2.9 @ 99/ £7.4.4
Midsummer 2/6 Stems 4/6 7.0
£7.11.4
Ded. Materials £1.0.7
Powder 18.–
Smith Cost 1.7.5
Drawing
Doctor 1.3 3.7.3 1/3 4.4.1

George Bailey & Prs Sinking Kemps Shaft 2.0.0 @ £16 £32 —
Do 0.4.1 @ £19 12.18.7
Midsummer 4/6 Stems 5/3 9.9
£45.8.4
Ded. Materials £3.3.11
Powder 2.14.0
Smith Cost 7.1.10
Drawing 13.10.1
Doctor 2.3 21.12.1 2/3 23.16.3
Doctor 6/ £63.16.1

19

Poldice Cost for June 1806

Anthony Lean & Prs driving the 48 fm level North of Kitty Billings 2.0.0 @ £6.6.0 £12.12.0
Ditto 1.0.0 7.—
Ditto 0.3.4 @ £4.17 2.15.0
Midsummer 2/6 Stems 26/9 1.9.3
£23.16.3
Ded. Materials £2.4.6
Powder 15.–
Smith Cost 17.3
Drawing
Doctor 1.3 3.18.0 1/3 19.18.3

Thomas Perkins & Prs driving the 72 fm level South of Grose's Shaft 1.0.0 £8.10.0
Ditto 2.0.3 @ £10 20.8.4
Midsummer 2/6 Stems 13/ 15.6
£29.13.10
Ded. Materials £1.7.5
Powder 1.17.6
Smith Cost 13.8
Drawing 2.2.6
Doctor 1.3 6.2.4 1/3 23.11.6

Anthony Sandoe & Prs driving the 60 fm level West of Rogers 2.0.0 @ £12 £24.0.0
Midsummer 2/6 Stems 12.6
£24.12.6
Ded. Materials £1.0.7
Powder 2.8.0
Smith Cost 1.16.–
Drawing 3.6
Doctor 1.3 5.9.4 1/3 19.3.2

Collan Williams & Prs driv. the 12 fm level West of Daws 3.1.6 @ £5 £16.5.–
Midsummer 2/6 Stems 10/9 13.3
£16.18.3
Ded. Materials £1.17.8
Powder 12
Smith Cost 13.10
Drawing 13.9
Doctor 1.3 3.18.6 1/3 12.19.9
Doctor 5/ £75.12.8

20

It has become a commonplace that the cost-book system had inherent weaknesses that left mines ill-prepared for short-term cashflow problems. The fact that most cost-book companies paid out profits at the end of every one, two or three month period – leaving nothing in the kitty for important capital schemes or for the payment of wages if ore sales were poor that month – appears, on the surface, to support that belief. However, these oft-repeated facts tell only half the story. It is also a fact that large, well-run mines, like Dolcoath and East Wheal Crofty in the 1830s and 1840s, paid out dividends that reflected the profits but kept enough back for the continued day-to-day expenses the purser expected. It is also a fact that mines such as South Wheal Crofty, Wheal Uny and Wheal Agar – which lost heavily for decade after decade in the 19th century and rarely looked like breaking even, never mind making a profit – kept going, with the principal adventurers happily (for the most part) paying their regular 'calls'.

Why was this? It was because of the inherent strength of the cost-book system as operated in Cornish mines. Most shares were held by 'in adventurers' – that is shareholders or adventurers who had an interest in the mine other than just their shares. Adventurers invariably included the mineral lords, who were paid their dues on ore raised whether the mine was profitable or not. These men were usually prominent supporters of the poor rate as well as landlords to most of the workforce. The principal adventurers always included suppliers of important services and gear to the mine. Coal, timber, gunpowder, fuse, engines, machinery, candles, picks and shovels were supplied by these adventurers, as well as the services of carpenters, coopers, engineers and blacksmiths. If we take just one example from the end of the 19th century we can see why a loss-making mine was allowed to soldier on with the blessing of the 'in adventurers' for over 50 years. South Wheal Crofty was divided into more than 6000 shares in 1891, the largest shareholders being the Williams family of Scorrier. On 9 July 1891 the Williams paid out almost £300 in 'calls' for the previous four loss-making months to keep the mine going. A considerable sum in 1891. However, during that four-month period they sold £1140 15s 6d worth of coal to the mine. Likewise, Holman Brothers paid out £14 in calls but received £134 for gear and services. Bartles Foundry paid £2 9s 6d but received £11 16s 11d. Harveys paid out £88 16s 6d and received back £135 4s 9d. Even the smallest 'in adventurers', such as local butchers and hoteliers, benefited by supplying the mine with food, drink and hospitality.[19]

The strength of the cost-book system was that the important adventurers had an interest in seeing the mine survive. The other strength – one that is rarely acknowledged nowadays – was that the mine formed a community that was connected vertically and laterally, so that everyone from the mineral lord down to the humblest engineman or miner felt part of that community, had a shared loyalty to it and wanted it to succeed.

OPPOSITE
Pages from the cost book of Poldice Mine, Gwennap, for June 1806. Eight tutworkers' accounts are entered with their earnings at so much a fathom and their deductions. The names of some of the men are still common in Gwennap Parish – Spargo, Lean and Williams.

East Wheal Lovell Mine, Wendron, cost and pay book. In the Cornwall Record Office and the Courtney Library, Truro, there are literally hundreds of cost books dating from the early 18th century to the 20th century. They contain detailed information on the activities of many thousands of miners over two centuries.

been resented, and tributers also felt they were being paid less for their work than hitherto. The normal one- or two-month delay in tribute payments was a further cause of agitation. The attitude of miners who returned from abroad could be paraphrased as: 'We send the ore up in one skip and you send our money down in the next!' Many felt the need for protection through a miners' union, but although attempts were made to organise workers, with occasional short-term success, in general such things had to wait until the changed conditions that followed the Great War. However, notwithstanding the absence of unions, the tin mines were to see a series of strikes by miners and surface workers, many of which were successful. Large groups of angry mine workers gathered in St Just, Camborne, Caradon and the Minions to demand an improvement in working conditions, a four-week month, improved cover by mine doctors, and a larger share in the profits when metal prices were high.[20]

Things were sometimes made worse by a shortage of skilled miners as thousands went abroad to seek their fortunes, and Dolcoath even threatened to replace those it lost with 500 Irishmen. The Irish had been pouring into Camborne since the potato famine and were for the most part treated with sympathy and kindness, but as competition for work increased when mines closed, resentment against the newcomers began. In the 1890s another depressing feature of resentment was seen as mine after mine experienced sabotage of one kind or another. Pitwork and pumps were damaged at Dolcoath and Killifreth; East Pool suffered attacks on her underground dams; and at South Frances, Wheal Basset and West Basset plant was vandalised and arson committed. Expensive engines were damaged or destroyed, and at Wheal Basset the counthouse was burnt down. Perhaps justifiably, the authorities tended to blame returned miners for these attacks as it was they who appeared to be at the forefront of the general agitation.[21]

Greed and fraud at the top

Another potentially more serious problem manifested itself in 1883. The first indications were seen in 1877, when G.L. Basset sold his shares in Dolcoath while retaining the high dues rate of 1/15th. In 1883 Basset demanded payment of a 'fine' of £40,000 to renew the lease, which was to run out in 1887. Twenty years before the Duke of Bedford had outraged the mining fraternity when he made such a demand on Devon Great Consols, but Basset's move seemed like sheer, unadulterated greed – and he was asking twice as much as the duke had. Dolcoath's shares collapsed, with £60,000 taken off her value in less than a fortnight. If the adventurers could have 'picked the eyes' out of the mine in the four remaining years of their lease they would had done so, to the loss of everyone including the Bassets. However, as this was impossible, despite the 'ultimatum' nature of Basset's demand they negotiated the 'fine' down to £25,000 and soldiered on. The big worry was that his action could encourage others to do the same, but, fortunately for the mining community, Lord Robartes, amid general approval, renewed the Tincroft lease with a sliding scale of dues payments based on the current metal price. Robartes made no demand for payment of a 'fine' to renew the lease.[22]

As if Dolcoath's problems were not enough, just as Basset was making his excessive demand the mine was hit by another confidence-sapping crisis. In May 1883 J.T. Mayne, a clerk in the account house, was found to have sold 203 non-existent shares for £12,000. John Mayne was sent to prison for seven years and the mine was presented with a large legal bill for his prosecution. Underground the flagship of Cornish mines progressed from strength to strength, but in the office and boardroom the ship looked as though it might sink at any moment.[23]

The century closes with restructuring and amalgamation

The answer to Cornwall's long-term problems appeared to lie in new capital investment, company restructuring, reorganisation of management, and modernisation and re-equipping of mine plant. During the 1890s and early years of the 20th century the most successful surviving mines were restructured as limited liability companies and the system that had served Cornish mines so well for centuries was put aside. Dolcoath, Basset, East Pool & Agar, South Crofty and Geevor all abandoned their old cost-book structures and became limited companies, with several seeking the remedy which had worked during the crisis years at the end of the 18th century – amalgamation. South Crofty joined with New Cooks Kitchen, East Pool linked up with her neighbour Wheal Agar, Carn Brea brought Cooks Kitchen and Tincroft into her already extensive operation, and the Basset Mines group was created from Wheal Basset, North Basset, West Basset, South Frances and West Frances. Money to finance these groups came from successful Cornishmen returning from South Africa and Australia, together with cash from London entrepreneurs and traditional Cornish mining families like the Williams. A host of Cornish businesses also invested heavily in these new mines, as they had when they were cost-book companies. With sound capital and constantly improving technology, the period of continuing crisis ended with an air of optimism among Cornwall's mining men.[24]

CHAPTER NINE

THE END OF AN ERA

20th century (1896–1998)

OPPOSITE
Scene from Trevaunance beach, St Agnes, in 1910. Wheal Friendly mine is on the horizon, dominating the hillside, which was covered by old mine workings.

IN THE 1820s, while briefly visiting Wheal Vor tin mine, Sir Stamford Raffles, founder of British influence in Malaya, said to one of the Bolitho family: 'You will live to see the day ... when the mines of Banca (Bangka) will eat up the mines of Cornwall.' For the next 60 years this threat waxed and waned until by the middle of the 1890s it became a reality. Malayan tin production boomed, largely financed by London money, while the price paid for expensively produced Cornish tin declined. Exacerbated by the abolition of silver in European currency, which caused a major drop in the value of that metal, the profitability of the Malayan mines increased dramatically. The reason for this was that in Malaya the workers were paid in silver but the mines themselves were paid for their tin in gold. The *Mining Journal* commented that 'Straits tin producers got the equivalent in silver of £130 per ton, the Cornish tin producers for the same tin £60 in gold currency'. Tin production from the easily worked Malayan alluvial beds went up from 26,000 tons in 1890 to 47,000 tons in 1894. To make matters worse, enormous quantities were beginning to come out of the Bolivian tin mines, 15,000 feet up in the Andes. Fortunately for Cornwall the altitude, climate, scarcity of labour and sheer remoteness of these mines kept their costs relatively high. Nevertheless it was the vast tonnage of tin available from abroad, together with the willingness and even eagerness of London investors to put their money into these foreign enterprises rather than the old-fashioned and struggling Cornish mines, that created the crisis of the mid-1890s.[1]

Reshaping of the industry in the early 20th century

Despite the gloom, the restructured mining companies that ran Dolcoath, Basset, Carn Brea and East Pool continued to plan for a better future. Basset Mines, which formed such an extensive group of mines along the Great Flat Lode, was formed into a limited company in 1896, with Frank Oates at the helm and Michael Williams and the Bolitho family in support as shareholders. Oates had made his money from Kimberly diamonds and was keen to invest it in his native Cornwall. The Bolithos were bankers with extensive interests in tin smelting in the county. Michael Williams, chairman of Dolcoath, was a merchant who also had shares in several other mines. Between 1904 and 1911 Geevor tin mine was also restructured as a limited company under the ownership of a company led by Oliver Wethered, who

Typical underground scene at the end of the 19th century – whilst three men drill, another climbs up the ladderway to the next level. Imagine the scene with only candles to see by.

brought in money from West Australian Gold Fields Ltd. Geevor was to work successfully for most of the 20th century. South Wheal Crofty, which had taken over the sett of New Cooks Kitchen in 1896, became a limited company in 1906. Crofty formed the main part of the Cornish Consolidated Tin Mining Co. Ltd., which included Botallack, Wheal Sisters, Phoenix (Caradon) and Clitters mine. The company had its own assay laboratory at Tuckingmill. Its money came mostly from London-based businessmen and it was to operate profitably for most of the 20th century. Carn Brea, Tincroft and Cooks Kitchen joined forces in 1896 and became a limited company in 1900, with the main shares taken up by Lord Robartes and the Bassets. It was an enormous concern that operated inefficiently as several disconnected workings. The mine struggled on until 1913 when its principal workings closed down, leaving the more compact and profitable part of the mine at Tincroft to continue. Tincroft closed in 1921, but the sett continued to be worked informally, with tributers working the levels that were kept dry by South Crofty's pumping engines. In the 1930s the northern part of the mine was worked from Crofty's Robinsons Shaft, and eventually was absorbed into South Crofty. Dolcoath joined the limited liability companies in 1895, when, with Michael Williams as chairman and £100,000 of subscribed capital, the new vertical 3000-ft (909-m) Williams Shaft was started. During the next quarter of a century Dolcoath made a profit of £697,244, and until the Great War broke out the old mine looked as sound as a bell.[2]

To many observers the Cornish mining industry looked in better shape than it really was. Scores of small, inefficient mines might have gone to the wall, but most of the larger mines had obtained new capital, more modern gear and better management. All, however, needed a good tin metal price, and this remained dependent on external factors. The vast tonnages from Malaya and Bolivia remained a problem, but with tin-plating requiring an increasing amount of the metal, demand was gradually growing. The need for tin-plated petrol cans for the rapidly expanding petroleum industry was a factor that undoubtedly contributed to the rise in the tin price after the turn of the century. From a low of £64 a ton in 1896 the price rose to £181 in 1906. The larger mines heaved a sigh of relief and recently closed mines looked to reopen. In St Just, St Agnes, Wendron, Illogan, Gwennap and several other mining parishes formerly abandoned mines began to buzz with activity. Even tiny mines found capital from London, and the invested money gave the period the epithet 'the electric boom' as modern machinery was installed and new electricity companies were set up to supply power. Levant, Dolcoath, St Ives Consols, East Pool and South Crofty all took advantage of its availability. The first pump to be powered by electricity in a Cornish mine was installed at Tywarnhayle, Porthtowan, in 1909.[3]

Following a high in 1906 the tin price settled down a little, but the several small mines continued to spend money on modern concentrating plant in preparation for

Cooks Kitchen Mine, Illogan, in 1893. The miners were sinking Chapples (Engine) Shaft below the 406fm level. The bar to which the drilling machine is attached is fixed horizontally across the shaft and the bottom of the bucket lift of the Cornish pump is on the right of the picture.

Cooks Kitchen Mine, Illogan, in 1893. The scene on the 406fm level above the shaft sinkers on the previous page. A kibble of rock has been hoisted and is about to be tipped into the waiting wagon. Note the compressed air-powered Holman winch for hoisting the kibbles and the pitman working on the pump.

reopening and larger mines, like South Crofty, Grenville, East Pool and Basset, installed improved machinery both on surface and underground. In 1912 there was another surge in the tin price, which reached £210 a ton, sparking another mini-boom. Killifreth, which had closed in 1897, purchased a second-hand 85-inch pumping engine as part of its ambitious plans for reworking. Wheal Peevor at Redruth, Droskyn & Ramoth at Perranporth, Wheal Hampton at Marazion and Garlidna and Boswin in Wendron all restarted in the hope of benefiting from the rising tin price.[5]

The First World War and its aftermath

The boom ended, as did so many things, with the outbreak of the Great War in 1914. The tin price fell back to £151 a ton, the best young men went away to war, materials and gear were in short supply or impossible to obtain, and mines such as Killifreth, Boscaswell, Condurrow, Wheal Jane, Wheal Coates, Wheal Vor and Wheal Peevor quickly closed. Carn Brea mine had already closed most of its underground operations, Levant – the greatest mine in the far west – was struggling, and even the mighty Dolcoath was showing signs of exhaustion. Dolcoath had lost more men to the armed forces than she could afford, and very soon men were being taken off development work to concentrate on production. This 'picking the eyes' out of the mine was eventually to be disastrous, for when the War was over no new tin ground had been uncovered for several years, the plant was unmaintained and in poor condition and there were insufficient cash reserves to finance a recovery.

As if the low tin price and high costs were not enough, a new problem asserted itself towards the end of the Great War – industrial action by a newly unionised workforce. Levant mine had suffered from union activity over its low wages for some time before the Dockers Union, which represented the miners, called a strike in 1918. The strike quickly grew bitter, and intimidation and violence became features of the action as extra police were drafted in to protect the pumps and other essential plant. After several weeks the men's pay demands were met and work resumed. By that time, however, the mine was in a parlous state and the deeper levels beneath the sea were abandoned. Worse was to come. While the owners and management discussed reorganising and re-equipping the mine, Levant suffered a major disaster – the worst in a Cornish mine for over 60 years. The man engine collapsed, killing 31 men and boys and shocking the whole of the mining world. The following year the surface workers at South Crofty also went on strike, but this time it started when some workers refused to join the union. The company retaliated by withdrawing the monthly tin bonus, which further damaged morale. Early in 1920 there was a strike at Giew mine, and throughout the mining districts there was agitation for improved

Three miners including a young teenager emptying an end tipping wagon into an ore pass in Cooks Kitchen Mine in 1893. The chain stops the wagon falling into the pass.

IMPROVEMENTS IN PROCESSING TECHNOLOGY

During the 19th century tin ore crushing and concentrating techniques had improved somewhat gradually, with the slow introduction of steam power and the adoption of labour-saving methods in the mills. Towards the turn of the century, however, there was an upsurge of innovation and invention. The Wilfley shaking table, developed over several years by the American mine entrepreneur Arthur Redman Wilfley, and the Frue vanner, also a product of American ingenuity, made tin ore dressing far more efficient and economical. Cornish steam stamps were being replaced by Californian stamps and air-cushion stamps, some of the best of the latter being developed by Holman Brothers of Camborne. These were far faster and more efficient than the older types. Another significant development was the introduction of electric power, which was being applied at the more go-ahead mines. Geevor's supplier was the Cornwall Electric Power Co., and East Pool, Dolcoath and South Crofty took power from the Urban Electric Supply Co. Air-cushion stamps, Californian stamps, rock-breakers and a host of other pieces of machinery were powered by electricity, as were the pumps in some of the new mines. To solve the problem of contamination of the tin ore by wolframite – a tungsten mineral – magnetic separators were introduced at South Crofty and Clitters mine in the early 20th century. This had the added benefit of creating a new source of income for the struggling mines, for tungsten was a valuable commodity with a growing demand. Ball mills to release fine cassiterite trapped in the pulp were introduced at several mines. Some installed complete new dressing plants, and more of these were constructed under cover. Improved methods made the treatment of concentrates to remove arsenic more efficient. In the first half of the 19th century 'burning houses' – which merely burnt off unwanted arsenides and sulphides – were replaced by calciners. The efficient Brunton calciner not only removed the arsenic but recovered it as a highly marketable product. In the early 20th century a new method of separating arsenic was introduced that involved the use of froth flotation.[4]

wages. A new and unusual feature of these industrial disputes was the willingness by both sides to talk and negotiate. The mine managers met to discuss reasonable pay levels, and plans were drawn up to bring some sort of uniformity to the pay rates for particular jobs. The miners were experiencing extreme poverty as they struggled to manage on pre-war wages with post-war food prices. This poverty became a social problem which affected everyone from the richest landowner to the hardest mine manager. Approaches were made to the Board of Trade for help, but – despite generosity towards the colliery owners, who had the greater leverage on the government – nothing was done for Cornwall. Men of influence like Lord Falmouth sought government help to pay for specific schemes to give employment to unemployed and destitute miners, but little came of it. The legacy of bitterness left by this perceived abandonment lingered for years to come. The tremendous contribution made by Cornish miners to the wealth of the country over many centuries appeared to be ignored, as did the willing sacrifice made by Cornwall's young men in the trenches of Flanders such a short time before.[6]

OPPOSITE
Tywarnhale mine in 1909. This was the first electric pump to be used in a Cornish mine.

Lean years – the 1920s

If things looked bad in 1919 and 1920, they were to get even worse in 1921. Fewer than 20 mines survived the Great War, and by the spring of 1921 – with the tin price reduced to half of its 1920 figure – only Giew mine, a small section of the defunct St Ives Consols, remained at work. Basset had closed in 1917, unable to cope with oceans of water and huge coal bills, and Dolcoath had closed in 1920 as costs escalated and production dropped. Wheal Grenville, Tincroft and Wheal Kitty stopped their engines and closed shop. South Crofty, East Pool & Agar, Geevor, Levant, Wheal Busy and Tresavean all suspended operations. A few tin streamers continued to wash tailings and store their black tin, but for the most part what they produced was not worth selling. Optimistic 'free setters' or tributers worked above adit level in long-abandoned mines, but they too found it better to store their output and wait for better times. Never in the many hundred years of Cornish mining history had the industry experienced such a near total shutdown. In 1920, 3065 tons of black tin was produced, in 1921 it was down to 679 tons and in 1922 it was a mere 370 tons, most of it from the floors of suspended or closed mines.[7]

In May 1921 there was a serious incident at East Pool that could have had dreadful consequences had the mine not been suspended at the time. A major collapse of ground occurred at Engine and Michells shafts. If men had been working underground there would undoubtedly have been loss of life, but the collapse still caused a serious problem as it effectively removed a large area of rich tin ground from future production. It was fortuitous that a rich lode had been discovered in the northern part of East Pool during the Great War, so the company was able to turn

Three machinemen operating two Tuckingmill Foundry drilling machines on 320fm level at Wheal Agar Mine, Illogan, in 1895. A mine captain is watching keenly. The two machines were mounted on vertical steel bars alongside each other.

its attention to the development of Wheal Agar sett to exploit it. The new discovery, Rogers Lode, was to carry the hopes for East Pool's future and a new shaft, to be called Taylors Shaft, was sunk to exploit the lodes in the northern part of the mine.[8]

In 1921, despite earlier disappointments, Cornish mine owners again began to approach the government. Under the provisions of the Trade Facilities Act, South Crofty asked for help to recover from the problems caused by the collapse of East Pool's workings and the consequent loss of her pumping capacity. Crofty's 80-inch Robinsons engine could not cope with the extra volume of water, so, with a £30,000 loan from the Treasury, the mine purchased the 90-inch engine on Wheal Grenville's Fortescue Shaft. Dolcoath, Tresavean and Geevor all followed South Crofty in asking for help. Geevor, which owned half of Levant's shares, received £10,000 from the government to get Levant into shape. Tresavean obtained a loan of £30,000 in 1922 and was able to continue until 1928. Dolcoath asked for financial help to move its operation to South Roskear mine but, although money was obtained, the shareholders had to raise a considerable sum to secure the government advance.[9]

Aided by these loans, Geevor and South Crofty quietly and profitably soldiered on through the 1920s. However, the big news of the decade concerned activity at New Dolcoath, which was centred on the workings at South Roskear mine and at East Pool, where development of the Wheal Agar section was supplemented by reopening of the abandoned South Tolgus mine. The principal part of Dolcoath's rebirth was to be the sinking of a new shaft at South Roskear, which it was hoped would intersect rich, unworked lodes at depth. Controversy and resentment followed the decision by New Dolcoath's board to give the contract to a Welsh company, Pigott & Son. The contract stated that the company would sink the shaft to a depth of 1680 ft (509 m) within two years, which it was thought would be a shaft-sinking record. The shaft was to be circular, 16 ft (4.8 m) in diameter and brick-lined throughout. It eventually reached a depth of 2000 ft (606 m). The headgear and various pieces of machinery, including pumps, were moved from the abandoned Williams Shaft to the new shaft. Although Pigott's was a Welsh company, much of its labour was local and some of the men brought to Cornwall for the job were Cornish miners who had gone to Wales to find work after the Great War. In September 1926 the mine had to raise another £50,000 to continue, followed by the same sum the following year. In August 1929 New Dolcoath stopped all mining, having failed both to find sufficient good-grade ore and to raise the cash it needed to continue. The new shaft had cost £300,000, but precious little ore of any value had been discovered. In March 1930, with Dolcoath owing nearly £80,000 in outstanding debt to the Treasury, a receiver was appointed by the Treasury solicitor to wind the company up.[10]

At East Pool, Taylors Shaft was down to 1215 ft (368 m) by May 1923, and more good tin lodes were being discovered to supplement the expected output from Rogers Lode. An abandoned 90-inch engine at Carn Brea mine was purchased for the new shaft, with the intention of supplementing it with electric pumps. In December 1927 the shaft had reached the 1600-ft (484-m) level, with a sump of 30 ft (9 m). By the end of the 1920s, however, concern was expressed at diminishing returns from the Rogers and Morering lodes as their payable lengths shortened at depth. East Pool's Tolgus project, mentioned above, involved driving a wide tunnel from the 255-fm level of the old Wheal Agar workings towards South Tolgus mine. By 1920 this 'Tolgus Tunnel' had been driven 1000 ft (303 m), at which point it cut Great Lode, which was enormously rich and very wide. Unfortunately, trouble with Wheal Agar's pumps and a large inflow of water brought work on the tunnel to a halt and a brick dam was installed to protect East Pool from this water. In 1923 East Pool's daughter company, Tolgus Mines Ltd, began to sink a new shaft on the long-abandoned Great South Tolgus mine. The shaft reached a depth of 1450 ft (439 m) in June 1925, and until 1926 East Pool remained optimistic about its Tolgus project, but in February 1928, with nothing of much of value found, it was abandoned.[11]

THE END OF AN ERA

With costs soaring after the end of the 1914–18 War, the price of tin was crucial to the survival of the remaining mines. In 1918 the average price had been £330 a ton, but by 1922, with the price down to £159, mines that had not already closed were about to. Thereafter, the tin price crept up to a high of £291 in 1926, and several mines looked to restart operations. Polhigey mine in Wendron reopened, and the western part of the ancient Great Work mine also began to operate as Wheal Reeth. A large number of mines cautiously began exploration work: Wheal Hampton at Marazion, Tregembo, Tindene and Wheal Squire on Mounts Bay, and Wheal Kitty at St Agnes were among the several new prospects. Once again miners were in demand. Until quite recently old men were still talking of cycling first to this mine and then to that to find work. One old miner said that his father worked at Wheal Jane in 1906, Wheal Peevor in 1915, Killifreth and Wheal Busy from 1919 to 1927, Tresavean until 1928 and then Kingsdown mine at Hewas Water. He finished his career in the survey office at South Crofty in the late 1940s.[12]

1930–45

However, the optimism did not last and the tin price fell steadily, until in 1931 it hit a low of £118 a ton. Once again it was external factors that decided the direction of Cornish mining, and after the disastrous Wall Street Crash of 1929 the county did

The entire workforce of Lady Gwendoline Mine, Breage, are gathered at the shafthead in 1937. Lady Gwendoline was one of a group of small mines that worked at Breage and Germoe in the 1920s and 1930s. It did not long survive the outbreak of the Second World War.

Painting by Terence Cuneo of the Holman Brothers boardroom with members of the Holman family discussing their future plans in about 1950. The Holmans were not only important manufacturers and exporters of mining machinery to all parts of the world but they were also significant mine adventurers in East Pool, South Crofty, Dolcoath and other Cornish mines.

not escape the effects of the world-wide slump that followed. With the active cooperation of the principal tin producers of Malaya, the Dutch East Indies, Bolivia and Nigeria, the International Tin Quota Scheme was set up in 1930 with the aim of restricting production until the tin price reached economic levels. The agreement was to last two years. Under the scheme almost all the Cornish mines agreed to suspend operations until the price reached £140 a ton. Only East Pool was allowed to continue production. At South Crofty and Geevor – the two mines which had survived the slump in a healthy state – no mining took place for the 12 months between October 1930 and October 1931. Astonishingly, during the dreadful years of constant upheaval between the end of the Great War and the tin crisis of 1930–31, South Crofty showed a loss in only two years of operation, 1921 and 1930. Geevor also turned in profits during this period, suffering losses in only three years.[13]

With the quota scheme terminated, the much reduced Cornish mining fraternity moved into the 1930s. For the mine owners these were to be good years, with Crofty and Geevor making continuous profits, but for the miners things were quite different. Before the Great War they had been relatively well off, with wages higher than most working-class men's, but after the war things changed for the worse. The 1920s had been hard for the miners, many of whom returned to 'the land fit for heroes' only to discover that they were treated as anything but heroes. Now in the 1930s conditions and wage rates were even worse. With world communism seen as a threat and many politicians and bosses regarding any demand for better wages and working conditions as symptoms of Bolshevism, there was a tendency in Cornwall to industrial polarisation: the men became more unionised and militant and the mine and factory owners became more reactionary. By the end of the 1930s things were becoming desperate, with many underground workers finding themselves forced to call on their parish to meet the cost of boots to work in. Geevor continued to be profitable and her miners remained quiet, but at South Crofty tensions came to a head in 1939 and almost the whole of the eastern side of the mine went on strike. Since the amalgamation at the turn of the century it had operated as two distinct mines: South Crofty, which was centred on Robinsons Shaft, and New Cooks Kitchen, which was centred on New Cooks Kitchen Vertical Shaft. Cooks men were marginally better paid than Crofty men and, as several militant Welsh miners were employed at Crofty, it was not long before the tensions led to a long drawn-out and violent strike. Police were drafted in from Wales, and as the situation deteriorated miners fought with the police and with their erstwhile mates who continued to work. Crofty men tried to blow up the Cooks headgear and succeeded in blasting one of the engine ponds, sending thousands of gallons of dirty water into some of their own homes. The strike ended only with the outbreak of the Second World War. Ironically, two of those who were arrested for violence during the strike became heroes at the Battle of Arnhem, which neither, unfortunately, survived. Their brother, Bill Gronnert, remained at Crofty as a shift boss until the 1960s.[14]

Scene at Killifreth Mine, Kenwyn, in the early years of the 20th century. The abandoned whim engine house is in the foreground and Hawkes' 85-inch engine is in the distance. In association with the neighbouring Wheal Busy Mine, Killifreth continued working until the late 1920s.

Elsewhere in the 1930s the pattern of opening and closing mines continued. The western section of Great Work was worked as Lady Gwendoline, Wheal Kitty operated on a fairly large scale, and at Mount Wellington a brave attempt was made to extend the surface workings underground. Despite these efforts, by the end of the 1930s only the three largest mines were still operating on an economic scale, and even East Pool was struggling to find sufficient good-grade ore. With the outbreak of war the government stepped in to support East Pool as the threat to tin supplies from the Far East was felt to be serious. After the Japanese occupied Malaya and the Dutch East Indies the search for metallic ore of any kind in Britain became desperate and every potential source of tin, tungsten and iron in Cornwall was examined. The price of tin crept up until it was almost at the level it had been at the end of the Great War, and despite wartime restrictions and difficulties of supply Geevor and South Crofty continued to do well, both making annual profits throughout the conflict. The biggest problem the mines faced was a shortage of labour, with skilled miners especially hard to find. So many young miners had volunteered for the forces before restrictions on their service came in that East Pool, Crofty and Geevor all struggled to maintain development and production. One solution was to recruit 'Bevan boys' to work underground, and several of these men were drafted into the mines. Ernest Bevan had set up a scheme by which men could be called up for service in the mines instead of the armed forces. This could not, of course, replace the skilled men who had been lost to the forces, but it it did increase the labour available. Then, in 1943, with so many Italian prisoners of war arriving in the country, all three mines applied to take on some to help operate the mines. East Pool took on 24 Italians, but as they were not allowed to work outside their camps after dark employing them was not without its problems.[15]

A scene of jubilation after three miners, trapped underground for two days by flooding, were rescued from the depths of Wheal Reeth Mine, Breage. The mine was flooded by water from Wheal Boys Mine in January 1937. It took 48 hours to pump the water out and rescue the three miners.

Post-war doldrums

When the war ended in 1945, new problems arose as government assistance ceased and new restrictions and regulations came into force with the arrival in power of the left-wing Labour government. East Pool closed and her 90-inch Cornish pump was kept at work by South Crofty to ensure that the mine's eastern workings were not overwhelmed by water. Both Crofty and Geevor suffered severe labour shortages as many men chose to work in other industries after their return from the war, and many others found employment in mines in Africa. Both mines recruited Poles and Italians to train as miners. Despite good-grade ore and a rising tin price, the two mines had fallen behind in technology and machinery. Carbide lamps had mostly replaced candles, but electric cap-lamps were rare. Primitive tramming persisted despite the limited introduction of rocker shovels (automatic loaders) and electric locos. Progress in introducing better and more efficient rock drills and air compressors had also been slow, and all these problems could be traced to a lack of finance.

Government restrictions and the lack of confidence of investors in speculative mining ventures in the UK were major reasons for the doldrums in which the industry found itself.[16]

1951–85: rising tin prices drive growth and modernisation

In the general election of 1951 the Labour government was defeated and the drive for public ownership and restrictions on industrial investment diminished. Almost immediately there was renewed interest in mining investment and applications to re-examine mining prospects grew in frequency as potential investors, mineral lords, the Ministry of Supply and the Ministry of Works all communicated enthusiastically. Mines like Mount Wellington that had closed before the war attracted particular interest, although this renewed interest extended from one end of the county to the other. Lucket mine, far over to the east by the River Tamar, prepared to reopen. At Geevor and South Crofty there were attempts to modernise, with electric pumps replacing the ancient Cornish steam pumps at Crofty and the purchase of more efficient rock drills. Other innovations at these mines were the use of tungsten carbide bits on drill steels and air-legs on machines. Despite these improvements the 1950s was a decade when both mines seemed to be waiting for something to happen, and they presented an old-fashioned and dilapidated picture.[17]

All this was to change rapidly as the continuing healthy price of tin metal began a further upturn. In 1951 the price per ton peaked at £1077 before settling down to figures of between £719 and £787 in the years 1952 to 1959. Thereafter the price rose until in 1964 it stood at £1240 a ton. Crofty and Geevor moved with confidence into expansion programmes, the former examining ground on the western side of the Great Crosscourse, at Dolcoath and the Roskears, and the latter planning a move into Levant and Boscaswell Downs. Crofty also looked at the rich abandoned workings in East Pool, Tincroft and Carn Brea with a view to exploiting the deep untapped tin there. Geevor explored Cligga mine on the cliffs at Perranporth. There was an air of confidence as drilling rigs appeared on hillsides all over the old mining parishes and investors began to pour money both into established mines and into exploration of the new prospects. Once again mining was a desirable occupation as memories of the bad old days in the 1920s and 1930s faded. Young men from many backgrounds queued to work underground, the mines were modernised and wages were among the best in the southwest. During the 1960s and 1970s Wheal Jane, Mount Wellington, Wheal Pendarves, Wheal Concord, Cligga mine, East Pool, South Roskear, Boscaswell and Levant were all reopened, either from the existing workings or as new mines. Almost all went into production – some very profitably, though one or two were failures; but between them they employed many hundreds of Cornish miners and put tens of millions of pounds into the Cornish

JOHN HUBERT TROUNSON (1905–87)

Involved in all this development and progress was Jack Trounson, who had enthusiastically called for investment in Cornish mines since the end of the Second World War. He left his position as chief surveyor at South Crofty in 1961 to become engineer at Camborne Tin Ltd, a company dedicated to finding new mine prospects in Cornwall and persuading mining companies to invest in them. Trounson was successful in that it was his constant advocacy of the future of Cornish mining which drew so much attention to Cornwall's potential. He was not always successful in individual cases, such as Wheal Vor, where much work over a long period failed to bring results, but for the most part Jack was at the forefront of the mining boom which began in 1962 and lasted until the tin crash in 1985. Jack Trounson died in July 1987 at the age of 82, still enthusiastic about Cornwall's prospects. Not long before his death he visited South Crofty mine with a bundle of plans under his arm to explain some aspect of the work which he thought needed doing on the County Adit, then being maintained by Crofty's sister mine, Wheal Jane.

economy. At Geevor there was a massive programme of expansion, which included plugging a hole in the sea bed above the long-abandoned Levant mine and driving an incline shaft for haulage for many hundreds of feet below the Atlantic Ocean into the old mine. A massive trackheader machine costing a quarter of a million pounds was used to drive the incline shaft down to 21 level, half-a-mile out from the cliffs. Several long crosscuts were driven into Boscaswell Downs mine and rich tin lodes were opened up. At Wheal Jane Consolidated Gold Fields Ltd poured vast amounts of money into the project to work the mine on a large scale. At Mount Wellington a Canadian company was no less ambitious, creating an expensive, state-of-the-art tin dressing plant. Wheal Pendarves and Wheal Concord were also opened up, the former as a small but modern mine and the latter as a more speculative venture.[18]

1985–98: the end of an industry

The Cornish mining boom came to an abrupt end in 1985 with the collapse of the tin price from over £10,000 to £3400. Without help no Cornish tin mine could survive in such a situation, and Wheal Concord and Wheal Pendarves, along with several small stream works, were immediately in trouble. Wheal Jane and Mount Wellington (which worked as one mine), together with South Crofty and Wheal Pendarves, were now all owned by Carnon Consolidated Mines. The extensive workings at Jane and the deep workings at Crofty meant that with the low tin price they desperately needed government help. Geevor was also in great need. They all looked to the Department of Trade and Industry for cash, and after a protracted period of negotiation Carnon was offered a package of £25,000,000 for Wheal Jane and Crofty. Geevor was also offered help, but the managing director thought it insufficient and turned it down. In April 1986 – in a blaze of publicity, emotion, recrimination and protest involving marches through London by miners from every Cornish mine – Geevor closed. Backed by loans and credit, Wheal Jane and South Crofty soldiered on for a few more years.[19]

For Geevor it was not quite the end as it reopened later in 1986 and again in 1987 to allow trammers to remove 44,000 tons of broken ore that had been left underground. In January 1988 another more ambitious plan was put into operation, and once again Geevor opened for mining. A few former employees returned, to be joined by several Crofty miners and one or two from Wheal Jane, Concord and Pendarves. Led enthusiastically by Kevin Williams, a former Geevor shiftboss, the miners were soon involved in a fully operational mine. Within months development was taking place on Hanging Wall Vein, stoping was being carried out on Coronation, Treglown, Whiskey Central and Sims lodes, and from every point of the compass trammers were moving ore to the ore pass at Victory Shaft. After six months the mine was actually breaking even and there was an air of optimism despite low wages, a leaking dry and old gear. Unfortunately, the tin price continued its fall, and in February 1990 Geevor was again on care and maintenance, with all mining stopped.[20]

For the other survivors a new crisis came in February 1991, when the government withdrew its support of Wheal Jane and Crofty, effectively reneging on the agreements that had secured their future in 1986. As a result Wheal Jane closed in the summer of 1991. At Crofty it was decided after two days' closure that the mine would soldier on with a much reduced workforce to extract what ore could be easily removed and to rescue valuable machinery. Everyone received the same fixed wage of £4.24 an hour, from the managers to the lowliest underground worker, which meant that only those who could afford it remained. Two days after the temporary closure 87 men returned to Crofty, with a further 50 men at Wheal Jane mill and 30 men stripping out Wheal Jane's underground workings. By the end of 1991 Crofty's workforce had increased to 192 and there was a feeling that 'We've survived so far, perhaps we've seen the worst of the crisis!' The year saw a loss of £79,000, which, although bad, was not disastrous. In 1992 the workforce continued to rise and stop-

South Crofty Mine in the early 1990s. The junction of 380-fathom (fm) level at New Cooks Kitchen Shaft and the First Sub-Incline Shaft. Looking down the shaft the lights at 400fm level can be seen. The ore conveyor is seen on the right and the loading platform is alongside it. The First Sub-Incline goes down to the 420fm level, from where the Second Sub-Incline Shaft continues to the 470fm level.

ing and development were undertaken in the Roskear, North Pool and Dolcoath sections, with lodes such as No. 4 and No. 8 in Cooks section producing large tonnages. Since the late 1970s the mine had been deepened from the 380-fm level by means of incline shafts, and when the crisis came the shaft-sinkers were heading for the 445-fm level on the second of these. Despite the gloom surrounding the events of 1985, they had continued to sink the Second Sub-Incline Shaft, and in 1987 they reached 445 level and pushed on towards the 470-fm level, reaching it in November. Here the shaft-sinkers met bad ground conditions and time was taken up in excavating the shaft station and environs. Nevertheless, in April 1989 sinking recommenced, and in November they were 2841 ft (866 m) below surface, which was at the same horizon as the 550-fm level of Dolcoath mine, the deepest mine in Cornwall. It seems ironic that this marvellous achievement – spearheaded by Jimmy

Clemence, one of Cornwall's finest miners – should have come just as the ancient industry seemed to be in terminal decline.[21]

With Kevin Ross's appointment as manager of South Crofty in October 1992 there were immediate plans to secure the mine's future. In that year the tonnage hoisted rose to 168,000 tons, rising again in 1993 to 180,000 tons with an increase in the workforce to 231. But in 1994 there was crisis yet again as production remained static, the workforce increased and the mine turned in a loss of almost £4,500,000. Underground there was a desperate search for new sources of payable ore, and miners returned to areas that had been left behind when the mine was being deepened and new levels were opened up. Work recommenced in Tincroft section on Pryces Lode and on Dolcoath South Lode between the 290- and 340-fm levels. The western workings on Dolcoath South Lode had been driven under the old Camborne Consols mine, reaching a point beneath Trelowarren Street in Camborne. North Pool Zone, close to Pool School, was being developed and stoped, and the eastern workings on Pryces Lode had almost reached Tolskithy. Despite these frantic efforts to stay afloat the mine was sinking under the weight of debt.[22]

David Eldred mucking the bottom of the Second Sub-Incline Shaft, below the 470fm level, in November 1989. He is driving an Eimco 625 loader. Note the rock bolts and wire mesh in the roof, to enable the miners to work safely in the poor ground conditions. The sump was at a depth of 2841 ft (866 m) and is at the same horizon as 550fm level at Dolcoath's Williams Shaft.

1994 saw fresh hope as the management put together a rescue plan involving a share issue which they hoped would raise a minimum of £1,800,000. In the summer a tremendous publicity campaign was launched to attract investment from local, national and international sources. Shares were offered in minimum blocks of £200. The response from local people was tremendous – they requested shares worth over £500,000. Two foreign companies also came in with a combined investment of £650,000. One of them, Crew Natural Resources, took a controlling interest in Carnon Consolidated and the company name reverted to South Crofty Plc. With the support of Crew the mine once again began to modernise and replace worn-out machinery. Production improved and the workforce rose to 322, but production costs still remained higher than income from tin sales. In 1996 the old Robinsons Shaft was taken out of use after the whole operation had been moved to New Cooks Kitchen Shaft. Preparations were put in hand to refurbish Roskear Shaft and to deepen it from the 2000-ft (606 m) level to Crofty's 400-fm level. A new 46-ft (14-m) high headframe and winder were placed over the shaft, and Wayne Brown, one of Crofty's younger miners, was given the task of raising up to Roskear Shaft sump from the 400-fm level. Roskear Shaft, sunk by New Dolcoath between 1923 and 1926, is 18 ft (5.4 m) in diameter and brick-lined throughout. It was also Crofty's main western ventilation shaft.[23]

The year 1997 began optimistically but ended in despair. High grades and good production figures meant that early in the year South Crofty was at last breaking even, but then the tin price began to fall, and by the spring it was £1000 less than it was 12 months before. In June it had fallen to £3200 a ton and the mine's situation was grim. The announcement of a 'temporary suspension of mining activity on 470' (the 470-fm level) was merely the precursor for worse news, and within months the old mine – which had carried the flag for Cornish mining through crisis after crisis – was itself closed. On 6 March 1998 the last miner came to surface, and all mining in Cornwall had come to an end. The following day, amid emotional scenes, Crofty miners – accompanied by former Wheal Jane, Wellington, Concord, Geevor and Pendarves men and many hundreds of others whose families had worked in Cornwall's mines over the centuries – marched through Camborne and Redruth to Crofty. The marchers, their heads held high, were led by their respective town bands and as they processed through the streets people stood in their doorways and wept.[24]

For the mines this seemed to be the end of the story, a remarkable story that stretches back for perhaps 4000 years, but for the miners this was not the end as they continue to carry their unique skills to the four corners of the globe.

CHAPTER TEN

A TREASURE HOUSE OF MINERALS

OPPOSITE
Sample of azurite from South Caradon Mine, Liskeard. Azurite is a carbonate of copper and this specimen is typical of the beautiful examples found in the Caradon copper mines. It can be seen in Truro Museum.

THERE CAN BE FEW PLACES ON EARTH where such a rich variety of metallic minerals has been exploited as Cornwall. Gold and silver have been found, lead, zinc and iron have been widely exploited, and tungsten and arsenic proved valuable by-products of the Cornish tin and copper mining industry. Antimony, sulphur, manganese, uranium, cobalt, bismuth, nickel and molybdenum have also been produced in modest quantities. Following an analysis of metal production in the whole of the United Kingdom during the 70 years before the Great War, the authors of an Exeter University publication stated: 'Cornwall produced a wider range of minerals than any other district (it was the only source of some of the rarest minerals) and was the only one to see large-scale mining continuing down to recent times.' In South Crofty mine alone over 30 mineral species have been identified, some of which – like the three uranium minerals zippeite, johannite and coffinite – are rare or very rare.[1]

Precious metals – gold and silver

Since the dawn of time gold has been the most highly valued of all metals. Gold has always signified wealth and the most powerful rulers of antiquity prized its ownership above all other valuable things. Significantly, gold is a rare metal found in few locations and in relatively small quantities. However, despite its limited global distribution, this precious metal has been found in Cornwall, although not in significant quantities. Analysis of gold artefacts, formerly attributed to being of Irish origin, has indicated that the metal used might rather have come from Cornwall. Penhallurick says: 'Most British and Irish bronze age goldwork contains traces of tin … the implication is that the gold came from a tin-bearing region.' Since medieval times gold has been found in tin stream works in Cornwall, and in the 16th century Thomas Beare referred to the common occurrence of gold in the tin ground. Beare related an incident he witnessed in 1550 at Lostwithiel, where a 'certaine gentleman present … gathering out from the heap of tyn certaine glorious cornes affirmed them to be pure gold.' Beare then says that the gentleman showed him a ring made from gold found in tin streams in the area, and that there was another gold ring from the same source worn by another gentleman. Beare then wrote about the presence of gold in the tin ingots being sold at Bordeaux to merchants from Florence. Appar-

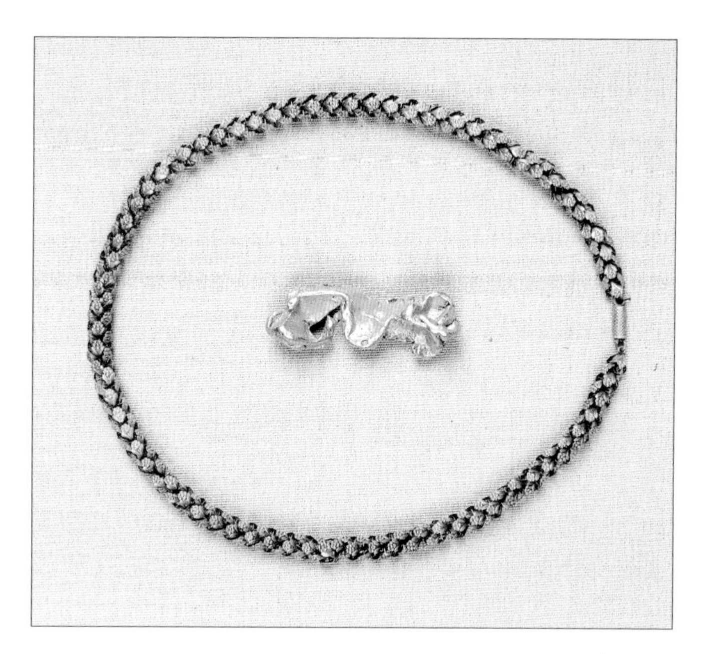

The gold nugget was found in Carnon Valley tin streams in 1808. The gold necklace was made from gold found in a tin stream below Ladock, near Truro, in 1802. They can both be seen in Truro Museum.

ently, the Italians were able to remove the gold from the tin ingots.[2]

Gold in small grains has been found in most alluvial tin ground in Cornwall and Devon. The valley below Ladock appears to have contained a significant amount of gold mixed with the cassiterite, and in 1802 an attractive necklace was made from it. A gold nugget discovered in the Carnon Valley tin streams in 1808 weighs 56 g (2 oz). This nugget and the necklace can still be seen in the County Museum in Truro. The nugget's source may have been the outcrop zones of the Gwennap copper lodes. Many of the gold fragments found there were attached to vein quartz. Analysis of chalcopyrite and pyrite from these lodes has revealed traces of gold. Little gold has been produced from the lode mines of Cornwall, but Levant Mine, St Just, did produce 4 oz (113 g) of gold from its copper ore, and at Treore Mine, near Wadebridge, gold was found in antimony sulphides. Native gold was also found near Treore in alluvial beds. Small amounts of gold have been found in crosscourses at Wheal Sparnon, Redruth, and at Wheal Vor, Breage. Despite excitement occasionally produced by unfounded rumours, gold has never been actively sought in Cornwall, and the only beneficiaries of its infrequent discovery have been the poor tin streamers who collected its tiny grains to sell to local gentry or goldsmiths when sufficient had been gathered. Carew reported in 1602: 'Tinners do also find little hopps of gold amongst their ore, which they keep in quills, and sell to the goldsmiths.' George Henwood said much the same in 1855.[3]

Silver has been found in several locations in Cornwall, under a variety of conditions. Records from the late 13th century show that silver was sought on the eastern side of the River Tamar at Bere Ferrers by the crown and its licensees. Miners were brought from Wales, the Mendips, the Pennines and parts of Cornwall to work the lead–silver lodes there, and at least four mines were operating in the district before 1300. These small, shallow mines were worked quite intensively for several decades before the ore began to run out and the technical difficulties of working at depth defeated the miners. They were revived in the 15th century, although silver production did not last despite the introduction of new pumping machinery. In 1788 silver was discovered at Wheal Mexico mine near St Agnes, and although the deposit was not extensive it does appear to mark the beginning of Cornish silver mining in recent centuries.

The commonest silver ore in Cornwall is galena, which may contain up to 170 oz (4.8 kg) of silver per ton of lead. The distribution of silver in galena lodes is very patchy, so that lodes can carry high values in one place and nothing within a short distance. The largest producer of silver in Cornwall was West Chiverton, which sold no less than 1,055,300 oz (over 29 tons) in the ten-year period between 1864 and 1873. The most important mine to work galena in Cornwall was undoubtedly East Wheal Rose, near Newlyn East, and although it was principally a lead mine, it did at times produce substantial quantities of silver. In 1852 the mine's lead ore contained 48,000 oz of silver, which was valued at about £12,000 – a useful supplement to the mine's income. In 1856 silver sales peaked at 53,280 oz, but after 1872 the mine sold no more silver. Among the biggest producers of silver from galena ore were the mines of Menheniot, near Liskeard. Wheal Mary Anne and Trelawney had significant quantities of silver in their ore, with the former reaching 67,210 oz in 1872 and the latter peaking at 51,516 oz in 1867. Nearly three-quarters of all lead mines in Cornwall produced some silver from their ore, and during the third quarter of the 19th century Cornwall was the most important silver mining county in Britain. Silver production peaked in 1869, when more than 300,000 oz (over 8 tons) was separated from lead at an average of nearly 47 oz (1.3 kg) per ton of lead.

Apart from the mines that worked principally for galena there were some which produced valuable quantities of silver from unexpected finds, as at Dolcoath and Botallack copper mines. In 1810 Dolcoath miners working at the 60-fm level on South Entral Lode discovered a substantial quantity of native silver associated with cobalt. This find, worth nearly £2000, was used by the Basset family to make silver plate. A far larger amount of Dolcoath silver was used in 1828 to make a decorative

East Wheal Rose lead and silver mine. This mine was the most important lead mine in Cornwall and was the scene, in July 1846, of the worst disaster in Cornish mining history. A cloudburst flooded the workings and 39 men and boys were drowned.

object weighing 40 lb (18 kg) and standing three ft (1 m) high. At Botallack in 1880 a large epergne made from the mine's own silver was presented to the purser, Richard Boyns. Silver was sometimes produced from lodes that were worked mainly for zinc, copper or tin. At Wheal Jane, for instance, the main tin lode consisted principally of sphalerite (a zinc mineral) along with some copper and silver, both of which were to some extent extracted. Wheal Jane's mineralisation appears to have been similar to that at Great Baddern mine immediately to the east, which also produced some silver. It is of interest that despite the relative insignificance of silver in the long history of Cornish mining, some was still being mined at Wheal Jane right up until the last decade of the 20th century.[4]

The base metals – lead and zinc

Lead was an important part of Cornwall's metalliferous mining industry during the 19th century, and for a couple of decades in the middle of the century the county was responsible for about 10 per cent of all United Kingdom production. As noted above, lead was mined together with silver on the Cornwall–Devon border from the late 13th century and well into the reign of Edward II. It was mined there again in the second half of the 15th century, although in both periods it was the silver that was sought. The Mines Royal also worked for lead in the second half of the 16th century, finding workable deposits at Treworthie near Perranporth and at Penrose near Porthleven. During the next two centuries lead was a useful by-product of many Cornish mines, although in areas like Nancekuke in Illogan it may have been sought as the primary metal.

The number of Cornish mines that have produced lead is probably well over 200, and during the period 1845 to 1913 about 170 mines recorded some production. However, as Dr Roger Burt has demonstrated, two-thirds of all output during those seven decades came from just five mines: East Wheal Rose and West Chiverton in the St Agnes–Newlyn East district, and Wheal Mary Anne, Wheal Trelawney and Herodsfoot mine in the Liskeard area. The most important of these, East Wheal Rose, began its period of high productivity in 1841, with 2030 tons of ore; this rose to a peak of 6885 tons in 1845 and thereafter gently declined to less than 60 tons in 1886. At its peak in 1845 East Wheal Rose produced two-thirds of all Cornwall's lead. The Menheniot mines of Trelawney and Mary Anne continued to be lead producers for a slightly longer period, drawing many miners from west Cornwall to the Liskeard area in mid-century and for a while creating a 'Wild West' atmosphere there. Undoubtedly, it was the relatively high silver content in the galena mined at Menheniot that helped to make these mines a success.

Lead mining has the dubious distinction of having suffered the worst recorded catastrophe in Cornish mining history. On 5 July 1846, when East Wheal Rose was at the peak of its career and employing over 1200 workers, disaster struck from an unexpected direction. A violent storm poured millions of gallons of water into the mine, drowning 38 men there and one in a neighbouring mine. The cloudburst appears to have been concentrated directly over the centre of the mine, which occupied a basin with its lower exit partially blocked by the tiny Metha bridge. Villages nearby saw hardly a drop of rain. The country rock at East Wheal Rose is relatively soft sedimentary material, is prone to water infiltration and had previously caused problems in wet weather. Two previous storms had caused partial flooding there. It was fortunate that the flood occurred when shifts were changing so that few men were in the ends and most were quite close to their access shafts. Despite the dreadful loss of life, there were also miraculous escapes as men and boys fought their way to the surface through torrents of water in the shafts. Predictably, some of the more extreme Methodists of the time saw the hand of God in all this, suggesting that it was the wickedness of many miners which had led to this judgement. 'Among so large a number of persons as were employed at East Wheal Rose – 1,260 – there were many of a very wicked and abandoned character ... the visitation is looked upon in the light of a judgement.' Not much Christian sympathy there, and very little was displayed in the outcome of the public subscription for the families either. Only £1718 was collected, with the adventurers contributing £1000. However, the flood did lead to improvements in the arrangements for surface drainage, with larger leats and better protection for previously unwalled shafts. This was fortunate, for the following year saw another cloudburst over the mine, and these added precautions saved the mine from another disaster.[5]

Zinc is widespread in Cornwall and has frequently proved a useful by-product to boost mine income. Cornish miners have used various words for sphalerite, the principal zinc ore mineral in the county – 'blende', 'black jack', 'jack' and 'wild lead' being the commonest. Like silver, zinc was frequently a by-product of lead mines, with as many as 50 of them producing saleable quantities. West Chiverton and Cargol mines were among the largest lead mines to produce significant amounts of zinc, and for both this by-product was crucial to their continued success. Indeed, until the 20th century West Chiverton was probably the county's largest producer of zinc. However, most of Cornwall's zinc came from mines other than lead producers, with significant amounts coming from copper, tin and iron mines. More than 50 in this group of mines produced zinc, the most successful of them lying in the Perranporth area. Great Retallack, Budnick Consols and Pencorse Consols mines were the top zinc producers in the 1850s and 1860s, during which time they accounted for nearly one-third of total United Kingdom output. During the 1870s West Chiverton and the Perranporth mines of Duchy and Peru enjoyed an increasing zinc output, which fortunately coincided with greater demand due to the introduction of better galvanising methods. Cornish zinc output remained significant through until the early 1880s, peaking at 7793 tons in 1881, which constituted over a quarter of the national total. Thereafter it declined quite steeply until by 1889 output was down to a mere 129 tons, a fraction of the national figure. Until the First World War zinc production came and went, with occasional output from places all over the mining districts, but until Wheal Jane began production in the 1970s zinc remained an insignificant metal in Cornwall. By the middle of the 1980s Wheal Jane was not only producing 1800 tons of tin-in-concentrate a year, 700 tons of copper and some silver, but also 8000 tons of zinc, which was more than the total Cornish output in the peak year of 1881.[6]

Iron and manganese

Occurrences of iron ore are widespread in Cornwall, although mining has largely been confined to one or two areas. The most important iron mining district lies in a band that stretches from St Austell near the south coast to Newquay on the north coast. Restormel Royal mine lies at the southeastern end of this band, with Rejerrah mine on the southwest. The iron mines around Whipsiddery and Newquay mark the northern end of this band. Mineral statistics from the years 1855 to 1913 indicate that some 80 Cornish mines sold iron ore during that time. Although iron ore was noted in the 17th century, little was exploited until the 19th century, and it was the middle of that century which witnessed the greatest output. However, even during the peak period from the 1850s to the 1870s Cornish output never reached 1 per cent of the British total. According to Burt, the largest annual tonnage was 55,150 tons in 1858, although Dines suggests a peak of 87,000 tons for one year in the period. By 1883 annual production was down to a mere 670 tons, and thereafter there were only occasional periods of production. Less than a dozen mines with an output of 10,000 tons each produced nearly three-quarters of Cornwall's total. Restormel Royal, Duchy & Peru, Trebisken and Wheal Ruby were among the county's largest producers. To put Cornish iron mining into perspective: despite the impressive-looking tonnage of iron ore mined, the total value for the whole of Cornwall for the 60 years up to the Great War was equal to the income of one medium-size Cornish tin mine over the same period.[7]

Manganese has never been an important metal in Cornwall, although some was mined in the 19th century in central and eastern areas of the county. Most manganese ores in the southwest are associated with culm measures and lie to the north and northeast of Callington. The iron mines north of St Austell also produced small amounts of manganese. In the second half of the 19th century Cornish production of manganese ore rose from 808 tons in 1867, peaked at 8671 tons in 1873 and then gradually fell away to 450 tons in 1881. Thereafter there was only occasional pro-

Abandoned mine buildings on the cliffs at Gravel Hill Mine, at the northern end of the Great Perran Iron Lode. This iron lode was one of the most productive in Cornwall.

duction until the 1890s. Although output was small, in 1880 Cornish production was almost 30 per cent of the national total. New uses for manganese were being found in the late 19th century, particularly as an alloy with steel and bronze. The oxide is used as an oxidising agent in various chemical processes, including to remove the greenish colour from glass caused by iron. Other uses for manganese include colouring glass (higher concentrations than those used to decolourise glass produce a violet hue), ceramics and paints.[8]

Antimony, arsenic and sulphur

Another element that has seen only tiny production is antimony. Although small mines where some antimony was found are spread across the north and east of the county, few produced it in measurable quantities, and total output for the 70 years before the Great War appears to have been no more than 35 tons, which was worth about £400. Bodannon mine at St Endillion, which worked in 1884, produced a mere four tons, which sold for £40. Pengenna mine, at St Kew, produced 15 tons in 1861, and Trebullett mine in Lezant raised 10 tons of antimony ore in 1891, which was valued at £250. Antimony is used with lead to form anti-friction alloys, as in white metal bearings, and in printers' type. It is also used in ceramic, glass, dye, cosmetic and medicine production.[9]

Arsenic, along with various sulphides, was for centuries an unwanted contaminant in the tin ores of Cornwall. Particularly after the late 15th century – when increasing amounts of lode ore were being prepared for smelting – these contaminants proved to be an expensive and difficult problem. From the end of the medieval period so-called 'burning houses' were employed literally to burn off the unwanted material, sending it up the chimney to foul the air and contaminate surrounding fields. In the early 19th century it was realised that arsenic could be used for a number of purposes, and by the late 1820s William Brunton had invented his highly efficient arsenic calciner. This not only removed the poisonous stuff from the tin ore but retained it as a merchantable commodity. Within a few decades sales of arsenic were proving a crucial support for many struggling Cornish tin and copper mines.

Although arsenic is found in many sulphide minerals, the only one of value in Cornwall is mispickel or arsenopyrite. The local name for this is 'mundic', which has tended to include both arsenopyrite and pyrite. Arsenopyrite can contain as much as 46 per cent arsenic and is worth working down to a content of about 20 per cent. Although arsenic has almost always been seen as a by-product in tin and copper mines, a few mines have worked for it specifically. Some have been reworked for arsenic that was left behind in old stopes, and this was especially true at Wheal Busy

Castle-an-Dinas Wolfram Mine, St Columb Major. The New Shaft is on the southern slopes of the hill, capped by an ancient earthwork that gives its name to the mine. Ore is fed by overhead ropeway to the mill at Old Shaft.

during and after the Great War and at Tincroft mine, which was accessed from South Crofty, in the 1930s. Both of these produced arsenic from long-abandoned copper stopes. During the last three decades of the 19th century huge tonnages of arsenic were being produced by the copper mines on the Tamar River, and Devon Great Consols, Gawton, Wheal Friendship and Okeltor were said to be producing half of the world's output at that time. The mines west of the Tamar produced more modest tonnages, but even those mines were large producers in the years between 1870 and 1885 – the peak year being 1885, when 3889 tons was produced.

Arsenic was crucial to the survival and profitability of South Crofty mine during the difficult first half of the 20th century. Sales of arsenic during the years 1900 to 1940 did not just keep South Crofty afloat – the mine's income from those sales was almost identical to its profit. South Crofty continued to produce arsenic from its seven Brunton calciners until the early 1950s, when a froth flotation method replaced them. Thereafter the material was once again viewed as a contaminant and disposed of.[10]

A constituent of pyrite is sulphur. Pyrite is a sulphide of iron and, like arsenopyrite, was called 'mundic' by the Cornish. It is found throughout the mining districts of Cornwall and is usually associated with other sulphide minerals. In gossan (the weathered outcrop of a lode) pyrite is generally decomposed, giving the appearance of a dirty, brown, rusty material. Underground its decomposition can produce heat, as in the mines on the Great Perran Iron Lode near Perranporth. Pyrite has been recovered economically in most Cornish mining districts, usually as a by-product. Only the mines in the far west appear to have lacked recoverable pyrite in their ores. The Callington mines in the east of the county and the Kea mines in the west have been the biggest producers, with Wheal Jane and West Wheal Jane producing probably the largest tonnage. These two mines between them sold over 60,000 tons of pyrite. The great copper mines on the Tamar River produced significant quantities, as did Prince of Wales mine in St Mellion. Pyrite can contain over 50 per cent sulphur, used to manufacture sulphuric acid, and during the third quarter of the 19th century Cornish production was important to that industry. Between 1850 and 1875 some 150,000 tons was produced in the county by about 70 different mines.[11]

Tungsten

An important mineral produced in Cornish mines from the end of the 19th century was wolframite. This is the main mineral from which tungsten is obtained. Wolframite is often found in association with cassiterite, the most common tin mineral, and as such was considered an unwanted and troublesome contaminant until the end of the 19th century, when reliable and economic methods were found to separate it. Cassiterite's density has enabled tin dressers to separate it from lighter gangue material relatively easily, and for centuries water has been used to wash away unwanted waste. Unfortunately, wolframite has a similar density to cassiterite, and this has made separation difficult. In the middle of the 19th century the Oxland process, which involved converting the wolframite to sodium tungstate, was introduced to deal with the problem. However, at the end of the 19th century a new, more efficient method – magnetic separation – was introduced. Tin dressers were able to use the slightly magnetic quality of wolframite to separate it from cassiterite. The system was introduced at Clitters mine and South Crofty in the early 20th century, and it remained in use at Crofty until the 1970s. In the 1980s and 1990s Mike Hallewell, the mill superintendent at Wheal Jane, experimented with a number of sophisticated methods to separate wolframite from a variety of complex ores both there and at South Crofty. He overcame the problems posed by Wheal Jane's ore, but found the presence of iron in Crofty's wolframite-bearing lodes more difficult to deal with.

Although wolframite yields were first recorded in Cornwall in 1858, it was to be the 20th century before there was regular production of the mineral, and it was warfare that stimulated demand and increased the interest of mine owners. Tungsten is used, among other things, to harden steel, and so the need to protect vehicles of war from enemy shells increased demand considerably. The Great War, the Second World War and the Korean War all saw demand increase.

Nearly 40 Cornish mines have produced some wolframite. It has tended to be found in the upper part of tin zones but, unlike cassiterite, it is rarely found in the adjacent wall rock. Although mostly found in the east-north-east-trending lodes, it has also been found in north–south lodes, as at Castle-an-Dinas near St Columb Major, the most important tungsten mine in Cornwall. Wolframite is frequently found in mineralised districts of Cornwall as rectangular-shaped black crystals embedded in quartz. At South Crofty it has been mined on both sides of the mine, but particularly in the two Complex Lodes. Very high values were found at East Pool near the Tolgus tunnel and at New Cooks Kitchen Pegmatite (Complex) Lode.[12]

Uranium

Among the rarer elements found in Cornwall at recoverable levels is uranium. Perhaps up to ten mines have produced some uranium ore, but the output was never significant. The two best-known producers of uranium were Wheal Trenwith at St Ives, and South Terras (Uranium Mines) near St Austell. The former raised 204 tons of ore between 1911 and 1913, and the latter 654 tons between 1890 and 1910. Edward Schiff, an American who first realised the commercial value of the uranium ore at Trenwith, also had interests in South Terras mine. He sent a box of pitchblende (uranium ore) to Mme Curie, who was able to isolate radium from it. Wheal

Wheal Trenwith Mine, St Ives. The mine produced over 200 tons of uranium ore between 1911 and 1913. The headgear of Victory Shaft can be seen, with some of the buildings which housed the concentration plant.

Edward, part of Wheal Owles, which lies on the cliffs at St Just, produced very small amounts of uranium ore in the 1870s, as did East Pool in Illogan. In 1954 the Atomic Energy Commission inspected South Crofty workings on No. 4 Lode near the granite–killas contact and found uranium but concluded that it was not worth working. Rare uranium ores were found in the 1990s on Dolcoath South Lode at South Crofty, but none had commercial value.[13]

Bismuth

Bismuth is another rare element, occurring in only four mines in Cornwall. It has been found as a sulphide in vughs (cavities in rock which may be filled with crystalline material) at Fowey Consols near St Blazey, and in crosscourses at East Pool in Illogan in association with cobalt and nickel ore. Some bismuth was produced at Dolcoath and Wheal Owles in St Just, each selling a mere 2 cwt in the 1870s. East Pool was the largest producer of bismuth, with a recorded output of 4 tons between 1872 and 1877. It was sold mixed with cobalt and nickel ore.[14]

Cobalt, nickel and molybdenum

Cobalt minerals have been identified in about a dozen mines in Cornwall, although ore has been raised in only half that number. Four mines appear to have had a recorded output of cobalt: Dolcoath, East Pool, Dowgas in St Stephen-in-Brannel and St Austell Consols. Dolcoath mine is said to have produced a few tons of cobalt ore, East Pool 4 1/2 tons, Dowgas 4 tons and St Austell Consols an impressive 250 tons, although the St Austell ore was mixed with nickel mineralisation.[15]

Closely associated with cobalt is nickel, which is found in the form of a sulphide and an arsenide. Eight Cornish mines have produced nickel, four in the centre of the county and four in the west. St Austell Consols produced over 12 tons of nickel ore, mixed with cobalt, between 1854 and 1861; Fowey Consols sold over 8 tons between 1858 and 1867, and East Pool mine raised and sold 4 tons of nickel ore in the second half of the 19th century.[16]

Molybdenum, titanium and plumbago are all rare minerals which have been discovered in Cornwall. Molybdenum is a sulphide only recorded to have been recovered from one mine: Drakewalls at Gunnislake. There are no records of the quantities of molybdenum raised there. Titanium, also known as manaccanite, due to it being first discovered in the parish of Manaccan in west Cornwall, was never sought commercially. Plumbago is believed to have been identified in several mines in Cornwall, but the extent to which it was worked is not known.[17]

Chalcophyllite from Wheal Gorland Mine, St Day. This beautiful specimen of green hexagonal plates of chalcophyllite forms a nest of crystals in red cuprite and white quartz. From Philip Rashleigh's Collection at Truro Museum.

Baryte and fluorspar

The spar minerals baryte and fluorspar are also found in Cornwall, although neither has been particularly important to mining there. Baryte (barite), or barium sulphate, is not common in the mining districts of the southwest, and what there is tends to be in Devon rather than Cornwall. In Cornwall it has been seen in lead- and zinc-bearing lodes, but not in commercial quantities. The north–south lodes of the Teign Valley have produced some baryte worth mining, especially at Bridford mine. Fluorspar, or calcium fluoride, occurs in some Cornish lead lodes and occasionally in copper lodes, where it is usually massive and can be either blue or green. South Crofty mine at Pool and West Wheal Damsel and Poldice mines in Gwennap have produced some fluorspar, but most production has come from the mines on the Devon side of the Tamar River – at Bere Ferrers, and in the Liskeard area at Wheal Trelawny, Wrey and Ludcott mines. Cornwall has also produced ochre from 18 mines, umber from six mines and fuller's earth from one mine between Newquay and Perranporth. None of these commodities has been significant to the Cornish economy, although some might have given slight, short-term benefit to the miners or mine owners.[18]

A mineral treasure house

Cornwall has long been, and remains, a rich hunting-ground for the mineral collector. Not only those minerals that have been mined commercially but also a great variety of rare mineral species are sought here by mineralogists. The wonderful collection put together by Philip Rashleigh in the 18th century can still be seen in the Royal Cornwall Museum in Truro. The Rashleigh collection has many specimens unique to the county or even to the particular mine where they were found. Eminent men and women like Sir Charles Russell and Marie Curie have visited Cornwall to search for specific minerals, and many books and learned articles have been written about the amazing variety of minerals found in the county. There is no question that Cornwall offers a cornucopia of rare and unusual minerals to the amateur and experienced collector alike.

ABOVE
This langite specimen comes from Levant Mine, St Just. It forms a crystalline crust of bright blue langite with green brochanite.

LEFT
A chalcotrite specimen from the Gwennap copper mines. This copper ore is known as hair copper because it has the form of very fine red hairs. Both specimens can be seen in Truro Museum.

CHAPTER ELEVEN

CHINA CLAY AND SLATE QUARRYING

OPPOSITE
Detail of painting by Dame Laura Knight of a china clay pit in 1912. Clay workers can be seen preparing some timbering.

TWO INDUSTRIES which have helped shape the Cornish landscape over the centuries are china clay and slate quarrying. The former is a relatively new industry, beginning in Cornwall a mere two and a half centuries ago, but the latter is truly ancient, having its origins back in the Bronze Age and possibly earlier. China clay was sought by Europeans for centuries before its presence was discovered in Cornwall. From the eighth century AD China had produced fine porcelain pottery by using a mixture of china clay and china stone. China clay is granite in which the feldspar crystals have decomposed into kaolin, a fine white clay. China stone, also known as petuntse, is granite which has been only partially altered by decomposition. After grinding the china stone, the Chinese potters mixed these two materials and then fired them at high temperature so that the stone vitrified and gave strength to the finished pottery. The porcelain that resulted was thin, delicate and translucent.

China clay

China clay was used for centuries before the 18th century for lining the inside of tin-smelting furnaces or blowing-houses, although its use in pottery was limited due the difficulty of separating it from the quartz sand with which it is associated and its lack of plasticity. In the Hensbarrow Downs area of Blackmoor stannary its widespread availability made it ideal for the tin blowers, as it undoubtedly did in the Breage and Penwith mining districts.

The first person to realise the potential of Cornish china clay and china stone was a Plymouth chemist called William Cookworthy. Born in Kingsbridge, Devon, in 1705, Cookworthy was a devout Quaker who trained as a chemist before establishing himself as a wholesaler to the trade. He was interested in geology – a very rudimentary science in the first half of the 18th century – and was particularly fascinated by the problem of how to manufacture porcelain successfully. He searched for the correct ingredients and even had samples of china clay sent to him from the American colonies. Cookworthy often travelled through Cornwall, examining the composition of rocks and learning about the minerals of the county. In about the year 1745, while examining ground that had been uncovered by mining and quarrying on Tregonning Hill, Breage, he found the china clay he had been searching for. Encour-

aged, he continued his search in other promising areas, and on Hensbarrow and at Carloggas he found much larger deposits of both china clay and china stone. In the years that followed Cookworthy used these two materials in experiments to develop a method of porcelain manufacture, and by 1768 he was sufficiently confident of success to take out a patent. He set up a pottery in Plymouth and started to make porcelain, eventually moving to a factory in Bristol. The business was not a commercial success due to his inability to solve the problems of large-scale manufacture, and so he passed his patent to his partner, Richard Champion, in 1774.

At this point the wealthy potters of Staffordshire took an interest in the new method of producing porcelain and set about getting in on the business. Led by Josiah Wedgewood, they opposed the renewal of the Cookworthy patent by Champion. After a court case in 1775 the patent was lost and the Staffordshire potters were able to obtain Cornish china clay and stone from local landowners, putting them in effective control of the industry from extraction of the raw materials to the manufacture of porcelain. This situation lasted from 1775 until the early years of the 19th century, when the Cornish took control of china clay and stone production. By 1850 these materials were being produced and sold to the potteries by a multitude of relatively small Cornish companies that operated modest quarries throughout the St Austell area. The industry was expanding rapidly and ripe for major investment by wealthy entrepreneurs.

OPPOSITE
East Caudledown Pit, near Bugle, in the heart of the clay country, in 1936. The engine house contains a 36-inch cylinder Cornish pumping engine built by Nicholls Williams in Tavistock.

The 19th century witnessed very rapid expansion of an industry which had started on such a modest scale only half a century earlier. In 1800 less than 2000 tons of clay was produced annually; by the middle of the century this had passed the 60,000-ton mark, and by 1900 more than 500,000 tons were being extracted and sold from Cornish pits every year. The largest producers were to the north and northwest of St Austell, with smaller quarries operating in West Penwith, at Towednack and St Just, and in Breage at Wheal Grey. Lee Moor in Devon was also a producer, as were quarries on Bodmin Moor. In the 1850s fewer than 90 china clay quarries were active, but by the outbreak of the Great War this had risen to almost 160. The organisation of these pits resembled that of the tin industry in medieval times, with every small, medium or large company operating independently of each other.

Charlestown Harbour in the late 19th century. Men are loading clay onto a ship from horse-drawn wagons.

They each had their own transport systems and their own harbour and shipping arrangements. They marketed their produce individually. The sett boundaries also resembled the bounds of tin works under the stannary system, with just as many restrictions and prohibitions on trespassing into neighbouring clay works. The industry was operating wastefully and inefficiently, with good clay ground lost due to the need to quarry, dress and store clay within the boundary of each sett. The need to dump stripped overburden within the sett only contributed to this inefficiency. The problems became worse as it became necessary to sink the pits deeper and the space needed to work them became more restricted.

Throughout the latter part of the 19th century the china clay industry was gradually modernised. Electricity was introduced in the 1880s and with it more efficient methods of extracting and drying the clay. The old-fashioned methods of working in the pits were being replaced by the use of high-pressure hoses to wash the clay from the sides. The West of England Company was the first to try this method in 1877, but it caused an angry reaction from the quarrymen, who sabotaged the hoses. During the next few decades such methods became standard and new high-speed pumps were introduced throughout the central clay districts, the water to feed them being taken from nearby flooded pits. With electricity also came centrifugal pumps to recover kaolin from the slurry produced by the new hydraulic mining method, and some of the large Cornish steam pumping engines in the pits were replaced.

By the time of the Great War most of Cornwall's china clay was being exported abroad, and when war broke out in 1914 the industry almost came to a standstill. As with the tin industry 20 years before, crisis meant amalgamation, and in 1919 three of the largest companies joined together to form English China Clays Ltd, universally known as ECC. This combination controlled half of Cornwall's output. Consolidation continued, and in 1932 ECC became English Clays, Lovering Pochin & Company (ECLP), with control of three-quarters of the industry. Keen to modernise further, in 1936 the new company built a large coal-fired power station at Drinnick, which supplied the whole of the central clay district with power. Overhead cable systems were installed in the pits to move clay, more efficient and economical methods of drying the clay were constantly being introduced, and transport between the pits and the ports was brought up to date. Between the wars the horse-drawn wagons were replaced by a multitude of small lorries, and as Pentewan closed due to silting and Charlestown declined, Par and Fowey became the main ports for exporting the clay.

Somewhat dwarfed by the great combines of ECC and then ECLP were a couple of independent clay producers, the Goonvean & Rostowrack Company and Steetley Burke. The former had pits at Goonvean, Rostowrack, Trelavour and Wheal Prosper, and the latter owned Greensplat and Bodelva pits. Like the giant clay company,

these smaller companies had to modernise, rationalise and progress to meet the challenges of the late 20th century.

Painting by Dame Laura Knight of men working in a china clay pit. China clay stone is trammed by workers along an elevated tramway before being tipped into larger wagons for transporting out of the pit.

The Second World War created problems similar to those caused by the Great War, resulting in the closure of several pits. As the industry was now effectively in the control of one large company the way was open to yet another phase of modernisation, with concentration on the larger, more efficient and profitable pits a priority. Rationalisation was now the order of the day and small, abandoned pits, derelict buildings and obsolete machinery became features of many parts of the landscape which such a short time before had been busy parts of the clay country.

Papermaking rather than porcelain was now the main market for Cornwall's china clay, and as a consequence production of china stone diminished. The years after the Second World War also saw a change in attitudes, and what might have been acceptable to the workforce before the conflict was no longer tolerated. Men demanded better wages, shorter hours and better working conditions. ECLP's

response was to increase mechanisation and reduce the labour force. Production reached 1,000,000 tons of clay in 1955, doubling to 2,000,000 tons in 1964. With the price of coal rising, the few remaining steam engines were phased out and electric pumps took their place. The cost of coal also meant that pan kilns were replaced by oil-fired rotary driers. These new methods led to the closing down of brick and tile works as their products were no longer required. The china stone mills had all stopped working by 1965, leading to local reductions in the number of those employed. Overall, however, the number of workers continued to rise, reaching a peak of some 6000 in 1970.

By the time the industry moved into the last quarter of the 20th century its whole appearance had altered. Men who had spent their lives working in the pits had seen vast changes. Where dozens of men had worked as teams to produce the clay in large quarries, now shifts of just three men supervised automatic monitors as the clay was hosed from the pit sides. Following the disaster at Aberfan, where an unstable pile of coal waste had buried a school with the loss of many lives, the traditional conical tips were made illegal and the waste had to be dealt with differently. New regulations on the quality of river water also meant that it was no longer possible to dispose of mica waste into the nearest stream, so huge lagoons were created to hold it and abandoned pits were backfilled.

OPPOSITE
Caudledown's abandoned engine house and clay pit in 1960. A typical scene in the years after the Second World War as smaller and uneconomic pits were closed and the sites abandoned.

The modern industry is highly mechanised and very productive. By the early 1990s the output of china clay stood at 3,000,000 tons a year, with most coming from ECLP's pits, although the smaller operators were still able to produce clay economically. In 1999 ECLP was taken over by a French company, IMERYS, which was established in 1880. Reductions in the workforce continued in the name of rationalisation and, theoretically at least, improved efficiency. In the middle of 2005 the number directly employed had dropped to 2130, with another 500 working as subcontractors, and plans were in place to reduce the workforce further to about 2000.

IMERYS currently produces about 2,300,000 tons of china clay a year, with its smaller neighbour, Goonvean, raising another 220,000 tons. As the industry moves into the 21st century it continues to dominate central Cornwall and remains one of the most important wealth creators and employers in the county, as it was for much of the 20th century.[1]

Cornish slate

An industry that undoubtedly rivals tin mining for antiquity is the quarrying of Cornish slate. Archaeologists have found an abundance of evidence throughout Cornwall for the use of slate by mankind as far back as the Stone Age. Pot covers or lids and spindle whorls used by Neolithic peoples more than 4000 years ago have been discovered in many places in Cornwall, and Bronze Age 'beaker folk' used large slabs of Delabole slate to cook on. From the Iron Age various slate artefacts have been discovered, including knives, pointers and piercers. These finds have come from sites at Newquay, Gwithian and Harlyn Bay. Evidence for the use of slate for roofing has been found from the Romano-British period. The villa at Magor Farm, Illogan, excavated in the 1930s, has left clear evidence of roofing slate, as have constructions dating from the fifth and sixth centuries AD.[2]

The medieval period has left ample evidence of the importance of the Cornish slate industry, and Delabole has been identified as an important source of the valuable material. Cornish slate was used for the roofs of the buildings at the royal silver–lead mines at Bere Ferrers, which were being worked in the 1290s. The records show that in 1296 23,000 'sclattes' (slates) were used to roof miners' dwellings at Martinstowe, and another 10,000 were used for the same purpose at Hassal, which might be at Combe Martin. In 1314 Cornish 'sclattes' were used to repair the roof of Winchester Castle. These slates were probably brought by sea to Southampton and thence inland to Winchester. In 1343 Delabole slates were used for the roof of Restormel Castle. We know that 19,500 were brought via the Fowey River to either Lostwithiel or Restormel, and another 85,500 from an unidentified quarry called Bodmatgan. As the latter were almost half the cost of the others – at 6d per 1000 as opposed to 11d – it is probable that Bodmatgan was close by, transportation of the slates being the greater part of their cost.[3]

The five quarries at Delabole, now joined up into one enormous pit, have been worked continuously since the early 16th century. Lying two miles from the sea, the present-day quarry is 500 ft (151 m) deep and well over a mile in circumference. The rim is 600 feet above sea level, so the bottom of the pit is 100 feet above sea level. Richard Carew, who visited Delabole at the end of the 16th century, lauded the quality of the slate there: 'In substance thin, in colour fair; weight light, in lasting long and generally carrieth good regard.'[4]

It was during the middle years of the 18th century that the ancient slate trade began to develop into the well-organised business it was to become. In 1750 Robert Bake, of an old Cornish family, was the sole proprietor, and under him and his family it became an efficient and profitable concern. His epitaph – appropriately inscribed on Delabole slate – can be seen on the wall of Lanteglos Church: 'In memory of Robert Bake, gent., of Delabole Quarry, in the parish of St Teath, who died the 10th day of October 1810, aged 84 years. As principal proprietor of the said quarry for upwards of sixty years. He justly bore the character of an honest man.' Although a devout member of the Church of England, this broad-minded man presented John Wesley with the land, building material and £40 to build a Wesleyan chapel at Delabole (Pengelly). The building was later used as a school. After his death the business passed to his grandson, also called Robert Bake, and he was to enlarge the original quarry and bring in improved working practices. He expanded the export trade to Europe and purchased three large vessels to carry the slate. His expansion plans included driving a half-mile long adit to drain the quarry more efficiently. This adit, which cost several thousand pounds, was completed in 1827, shortly before his death. Following his death there were several years of confusion over the quarry's ownership and its management, but in 1834 things moved forward again with the installation of the first steam engine at the works. This was used for haulage, and in the early 1840s there were in addition two horse whims for haulage and one large water-wheel for pumping water up to adit level.[5]

Quarrying, though not as dangerous as mining, is still a relatively dangerous occupation. Delabole has had its share of tragic accidents over the centuries, and four such accidents from the middle of the 19th century serve to illustrate the dangers inherent in the job. In 1843 six men were killed when they were carried away by the collapse of a huge piece of rock they were working on, most of them drowning in flooded old workings. Two years later a piece of ground fell into the quarry where a group of men were working 250 ft (75 m) below. Four were killed instantly. In 1847

Loading Delabole slate at Port Gaverne at the end of the 19th century. Note the women on board being handed the slate by the men on the horse-drawn wagons. The vessel is the ketch Surprise, which was built at Port Gaverne in 1879. It has been beached on a high tide to give maximum time to load.

Panoramic view of Delabole slate quarry in about 1900. The picture shows the south end of the quarry and the water wheels, which provided power to the workings. Note the tracked incline in the left foreground and the ropeway in the background.

there was a similar accident and again four men died. The worst accident, in 1869, killed 15 people, including a young woman and two children who were bringing food to the workers. An enormous area of ground weighing an estimated 10,000 tons subsided into the quarry, carrying the 12 men and three visitors with it.[6]

The hey-day of slate quarrying at Delabole was probably the few years prior to the Great War, when some 500 men worked there. In the 1980s, when Rio Tinto-Zinc (RTZ) took over the quarry, there was a workforce of about 80. Carnon Consolidated, RTZ's subsidiary, which managed the quarry, had ambitious expansion plans. Carnon proudly pointed to several prestigious buildings which had used Delabole slate in the 1980s, including the large new Marks & Spencer store in Truro and Truro Cathedral. Everywhere in Cornwall can be seen the distinctive blue-grey of the large slabs of Delabole slate. Old public houses and historic buildings display this beautiful stone, as do the many modern buildings where the slate has been used in various guises and colours. The current owners of the quarry, Mr and Mrs Hamilton, employ over 40 workers and sell directly to the trade and the public.[7]

Delabole has not been the sole source of Cornish slate. In the Liskeard area – particularly at Carnglaze in the parish of St Neot – slate has been quarried extensively.

The cave complex at Carnglaze is one of Cornwall's truly amazing sights. It bears witness to an industry that has long gone but which has left many folk memories among the people of St Neot and the Glyn Valley. The St Neot quarries produced slate for more than 200 years before finally closing in 1903. At its peak some 20 teams of quarrymen were working the slate, most of which went to customers in Plymouth. The last mine captain there, from 1880 until 1903, was Captain Edward Pascoe, whose grandson Alan has written a history of the quarries.

Stories are still recounted about how slate was brought down through the St Neot and Glyn valleys to the quays at St Winnow. In the absence of tramways or railways, pack-horses and mules had to be used. These hardy animals wended their way down through the valleys by way of Steppy Lane and Ley Village to Bodithiel Hill and then on to Middle Taphouse and St Winnow Quay on the Fowey River. The poor creatures then had to do the return journey up hill, this time with lime for the farmers. Tradition tells us that beneath this legal merchandise was often hidden contraband of wine and brandy smuggled past the revenue officers and up the Fowey River.[8]

It is heartening to know that this most ancient of Cornish industries continues to survive in the 21st century. Although much reduced and with far fewer employees than in the past, slate quarrying at Delabole continues, not just as an employer and bringer of wealth to a remote part of Cornwall but as a genuine link with Cornwall's past and a reminder that Cornishmen can still wrest a living from the hard, unyielding rock upon which the Duchy sits.

Quarrying slate at the foot of the cliffs at Tintagel, before the First World War. Note the man with the compressed-air drill in the background; the men in the foreground trim slate in the traditional way.

CHAPTER TWELVE

CORNISH MINERS AND MINING INTO THE FUTURE

OPPOSITE
Aerial view of one of Seacore's ships at the North Pole, where they were drilling deep into the sea-bed beneath the ice. The Cornish crew, including former Cornish miners, successfully carried out this operation where others had failed.

WITH THE DEMISE of active metal mining in Cornwall, it might appear that only china clay and Delabole slate quarrying survive out of Cornwall's once extensive extractive industries. However, that picture fails to tell the whole story. Not only have these industries survived into the 21st century, but several mining-related businesses also continue to work in the Duchy, and Cornish miners remain active both in their traditional work and on a host of civil engineering projects. Mining consultancies, mining contractors, suppliers of sophisticated mineral processing plant and laboratories for mineral analysis are all based in Cornwall and continue to supply the needs of the world's mining companies. Cornish miners trained underground at Geevor, Wheal Pendarves, Wheal Jane, Mount Wellington and South Crofty continue to practise their trade across the globe. In several countries of the former Soviet Union, in Ireland, Wales, Bulgaria, Romania, Albania, Greece, Turkey, Saudi Arabia, Sudan, Ghana, Nigeria, South Africa, Australia, New Zealand, Brazil, the USA and Canada – in almost any country where metallic minerals are extracted – Cornish miners will be found making their contribution. Cornish mining might be in repose, but the miners and their skills are still a significant part of the global search for and extraction of metals.

Consultancies, laboratories and contractors

When South Crofty, the last Cornish mine, closed on 6 March 1998, several of its ancillary businesses remained active. Crofty Consultancy was established in the late 1980s as the mine owners sought ways to help solve the many problems left by centuries of mining activity. Crofty Consultancy boasts the largest privately owned archive of mining records and plans in the southwest. Its mining geologists and engineers, led by Derek Morgan, are all graduates of the Camborne School of Mines, and several have been employed at South Crofty and Pendarves mines. Other firms had already been established to respond to the many mining-related questions raised by potential purchasers of property in the county. Cornwall Mining Services and South West Mining & Properties, both established by former miners and mine engineers, were already carrying out mining consultancy work by the early 1980s,

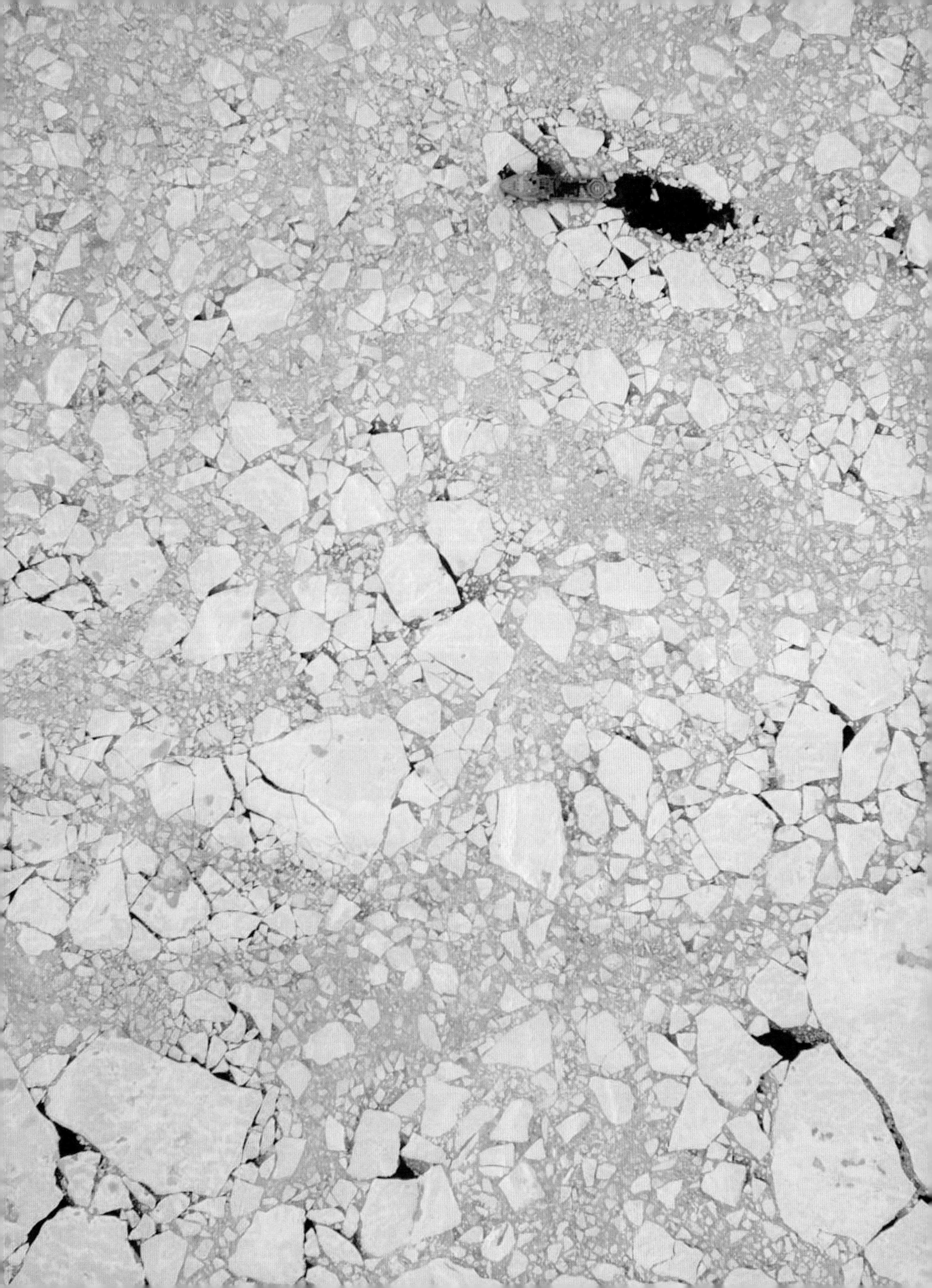

Geologists carrying out searches for mine reports in Crofty Consultancy's office at Wheal Jane. Reports on historical mining, site investigation work and soil and water sampling are constantly needed to ensure the safety of existing and planned buildings in the south-west of England.

and these businesses have been joined by others eager to offer advisory services relating to new legislation on property safety. Most employ former miners, engineers and geologists, as well as graduates of Camborne School of Mines. The work they do includes conveyancing mine searches, testing for mineral contamination of concrete ('mundic tests') and site investigation of buildings or building sites by drilling or trenching. Soil and water are also tested for contamination.

Wheal Jane Laboratory, run by Clifford Rice, is another example of continuity in the Cornish mining industry. Clifford, the chief chemist, and his assistant Keith Blunden were employed for many years to run the laboratory for RTZ and Carnon Consolidated before the company was taken over by Crew Natural Resources, a Canadian-based mining company. When Wheal Jane and South Crofty mines closed they continued to carry out analytical work for the company as well as contract work in several foreign countries. Clifford and Keith have helped set up and train operatives to use mineral processing plants in several places, from Camborne School of Mines to RTZ's mines in Spain and Portugal. Clifford has also pioneered techniques for the analysis of contaminated ground to establish whether the various poisonous materials may be absorbed by humans.

Another of South Crofty's businesses which continued to operate after the mine's demise is Carnon Contracting. This company, along with other similar businesses in Cornwall, has carried out a variety of jobs involving the use of miners and mining gear. Carnon was at the forefront of projects to set up geothermal energy units throughout the country, drilling holes hundreds of metres deep for the installation of ground-source heat pumps. The company has sunk shafts, driven tunnels, carried out exploration drilling and performed a host of other tasks in Wales, Ireland, Scotland, France, Portugal, Greenland, Ukraine, Sierra Leone, Greece, South West

Africa and Romania. Carnon Contracting employs a dozen men, mostly former miners, who learned their skills at Geevor, South Crofty and Wheal Jane mines. The workforces of other contracting companies are also largely made up of former miners and others who worked at the mines.

Drilling

Drillserve Ltd was set up in 1979 by Peter Sheppard, a graduate of Camborne School of Mines, who had worked at Wheal Jane during its initial pre-opening period in the 1960s. The company employs nearly a dozen workers, most of whom had worked either at Holman Brothers (see below) or in the Cornish mining industry. Originally the firm was set up to serve the needs of local mines and quarries, but now it supplies, builds and operates specialist drilling rigs and other items of mining and civil engineering gear to companies all over the globe. Drillserve has supplied

Robbie Osborne operating a drill rig for Carnon Contracting. This type of rig is used for testing ground conditions and mineral exploration. Carnon Contracting have carried out drilling contracts above and below ground in many different countries around the world as well as throughout the United Kingdom.

mining equipment to Antigua, Armenia, India, China, Ghana and Norway. In the UK the firm has supplied machinery to barite mines, salt mines, Bath stone mines and to several Cornish mines and quarries. Important contracts have been successfully completed at sites as different as Botallack mine, where remediation work was done, the Devonport Docks Submarine Refit Centre and the Moorcroft Tunnel in Plymouth. Peter Sheppard's men have also completed contracts in the Channel Tunnel, the Channel Tunnel Rail Link and at Terminal 5 of Heathrow Airport. Perhaps the most distant and remote location for their expertise has been at Mount Alice and Mare Harbour on the Falkland Islands. The men at Drillserve are fitting examples of Cornishmen's willingness to go anywhere and carry out any task, no matter how remote the site, difficult the terrain or complicated the work.

Saxton Deep Drillers Ltd has been working in Cornwall since 1976 – although the largest part of its operation is in South Wales, where the company operates opencast coal mines. It was in South Wales that the company first started. Saxton currently employs between 60 and 70 workers, most of them in Wales. The Cornish side of the business employs several former Cornish miners, much of whose work is for IMERYS in the china clay industry, where they do the drilling in preparation for granite to be blasted. In civil engineering the company is much involved in piling and rock anchoring (stabilising rock slopes and walls by putting in deep bolts), and it also does some geothermal and exploration drilling. A considerable amount of Saxton's work in Devon and Cornwall is borehole drilling for potable water.

The owners of Seacore Ltd do not consider themselves as running a mining company, nor do they see themselves as part of the mining industry. However, many of their most skilled men are former miners and mine workers. Alan Pope and Donald Matthews, two of Seacore's most experienced drillers, were both trained as hard rock miners at South Crofty. Andy Seagar was a mine surveyor and John 'Jango' Laity was a mine electrician. A large part of the skill base at Seacore derives from mining and the drills they use are basically similar to the rigs found throughout the mining world. Although Seacore specialises in drilling into the sea-bed, the company also carries out work on dry land anywhere its expertise is needed. The now famous picture of the group of former Cornish miners at the North Pole after they became the first team ever to drill through the ice cap and into the sea-bed beneath it tells us everything about the unique skills of this firm's drillers, as before them engineers and drilling teams from several countries had attempted the task without success. Despite the fast-flowing tide between the ice and the sea-bed, which changed its direction of flow without warning, Alan Pope and his mates carried out the required drilling operation successfully. Within a few weeks they were off to the Antarctic to drill there. Seacore might not see itself as part of the mining industry, but its former miners continue to display the unique skills learned deep underground in Cornwall.

Three ships involved in the drilling operation by Seacore at the North Pole. The far ship and the one in the foreground are ice-breakers and the one in the middle, the Vidar Viking, has the specially designed drill rig mounted amid-ships.

Tunnelling, mining and civil engineering

South Western Mining & Tunnelling Ltd is a firm run by brothers Robin and Charlie Daniel. These two former South Crofty men began the company in 1984 and since then have been actively employed all over the UK and Ireland. In the company's first year they were involved with mining at Gwynfynydd Gold Mine in Wales, where one of the brothers was manager. They then drove a long drainage adit tunnel at Meldon Quarry in Devon before crossing the country to carry out safety work at an old chalk mine in Norwich. A chalk mine in Berkshire was converted by the firm into a wine store. Over the years several contracts for mining work have been completed in the gypsum mines of Sussex. For almost five years the Daniel brothers ran a large barite mine in County Tipperary in Southern Ireland. The firm has driven major tunnels in Somerset and South Wales using traditional Cornish drill and blast techniques, and it was heavily involved in several aspects of the 'clean sweep' project to modernise the sewage disposal system for west Cornwall. Here they drove tunnels, sank shafts, blasted out trenches where conventional methods had failed,

The tunnel driven below the Devon-Somerset border by Carnon Contracting for South West Water. This work was carried out by miners trained at South Crofty, Wheal Jane and Geevor mines.

mined a large pump room and carried out all rock removal work. These former Cornish mine workers have demonstrated their skills – learned first at South Crofty – in every type of mining and civil engineering situation from pure mining to general ground stability work on cliffs, in motorway construction and down old mine shafts. They continue to fly the flag for Cornish mining.

Mineral processing

Holman-Wilfley is a name that has been familiar in the mining world for well over a century. Holman Brothers of Camborne were engineers of national and international fame for more than 200 years, and Wilfley shaking tables have been known as the Rolls-Royces of shaking tables since the late 19th century. This deceptively simple piece of equipment, on which materials of different density are separated by shaking or vibration, has for decades been part of the mineral processing plant at many mines. Holman Brothers began making shaking tables over a century ago, and along with the great variety of mining machines and engines they made, these tables were exported all over the globe. The current managing director of the firm that makes Wilfley and Holman shaking tables is Mike Hallewell. Mike managed South Crofty's tin mill before taking over the running of Wheal Jane mill, introducing many innovations as he sought to improve the recovery of tin despite ores that were sometimes very complex. When both mines closed in 1998 he moved on to Holman-Wilfley. The business has expanded continuously since then, and now specially designed tables are developed and sold all over the mining world. The company employs nearly two dozen full- and part-time workers. No matter what the mineral or how complex the ore, Holman-Wilfley will design a table to concentrate it. As a consultancy and a supplier of sophisticated mineral processing equipment, Holman-Wilfley is involved all over eastern and western Europe, North and South America, the Middle East, the Balkans and throughout Africa. Indeed, there is not a mining field where Holman-Wilfley tables are not in use, and Mike travels constantly to advise on and direct the installation of this much sought-after equipment. The skilled men who manage and operate the company tend to have a mining background, some having formerly worked at Wheal Jane under Mike himself. Recently it joined up with SGS Lakefield Research, a world leader in the field of mine research.

A company with plans to reopen South Crofty

Baseresult, the current owners of South Crofty mine, display the kind of optimism that was once such a feature of Cornish mining. Throughout its long history miners and mine owners have pushed forward with their plans despite the least promising circumstances. With so many things stacked against them, it is encouraging that

Holman-Wilfley men working on small shaking tables specifically designed to concentrate fine gold. Holman-Wilfley plant, research and offices are based at Wheal Jane Mine, where several mining-related companies are located.

Baseresult's management and staff remain determined to succeed. Pressure from development agencies and regeneration organisations, hostility from local councillors and council officials, excessive demands from health and safety interests and local concern engendered and encouraged by a suspicious press have all failed to stifle the company's enthusiasm to achieve its aims. In its avowed long-term aim to resume mining at South Crofty, it does not share the pessimism felt by many observers due to the tin metal price or the cost of pumping out water and re-equipping and regrading the levels.

Despite this demoralising climate, Baseresult has continued its preparations to restart mining. A modern, diesel-powered scooptram has been purchased at a cost of £40,000 so that miners can clean out the Tuckingmill Decline and begin tunnelling. A new Alimak lift has been installed in Old Engine Shaft at a cost of some £200,000. This can carry up to 16 people between the surface and the 150-ft level. At the moment it is used for men and materials and for tourists who visit the workings as part of the company's strategy to defray some of the operating costs. Commenting on the significance of the lift, Kevin Williams, Baseresult's managing director, said: 'It represents an important milestone in the mine's progress toward full operational status, the Alimak will be used when New Cooks Kitchen Shaft has the suspended submersible pumps in it, that will start the de-watering process.'

OPPOSITE
Group of miners having their 'croust' (lunch) break on 380fm level at South Crofty Mine, in about 1990. The shift boss Raymond Thomas is on the right and Ted Wilson, a miner who spent many years stoping at the mine, sits at the far end.

Baseresult currently employs 17 full-time workers, two-thirds of whom are former mine workers. These men were all trained at Geevor, South Crofty or Wheal Jane and firmly believe that they are participating in a project that will ultimately see Cornish mining resume at their mine. It is to be hoped that they are right and that their optimism will be rewarded. Mining at South Crofty was carried on for almost 350 years, with only two or three short periods when work was suspended. It is sobering to think that the mine has now been idle for longer than at any time since the Restoration in 1660. No true Cornishman or woman could wish for such a situation to continue, and if Baseresult and its group of optimistic workers can succeed the whole of Cornwall will applaud them.

A long tradition continues

Cornwall's extractive industries continue to move forward into the 21st century. The china clay and Delabole slate businesses are examples of continuity, and both have a future in the economic and social life of Cornwall. Just as significantly, the many Cornish companies which continue to operate in the mining and civil engineering fields at home and abroad demonstrate the resilience of Cornwall's long traditions. But above all it is the character of the Cornish miner – who since the dawn of history has demonstrated that he is among the best in the world – which gives the Cornish people their pride.

Glossary

Adit Tunnel driven almost horizontally from low ground to allow natural drainage for mines on higher ground. Also gave access and ventilation to shallow workings as well as allowing miners to discover new lodes by driving crosscutting tunnels.
Adventurer Old name for a shareholder, as working the mine was considered an adventure.
Alluvium Debris including rock, sand and clay deposited in a river valley.
Alluvial tin Tin found in the alluvium of a river valley. It was washed by the action of water from the surface outcrop of tin lodes into the nearest river valley over the millennia.
Arsenopyrite An important arsenic ore. Also known as mispickel or mundic – chemical composition iron–arsenic sulphide.
Assay hatch Exploration pit dug to search for a lode outcrop. If the lode was found the hatch became an access shaft. *See* Costean pit.
Assaying Method used to determine metal content of ore. *See* Coin.
Back Can refer to the visible part of the lode at outcrop or the roof of the mine when working underground.
Bailiff The man responsible for the day-to-day running of a stannary district. He was answerable to the Steward of the stannary.
Ball mill Cylindrical rotating mill in which ore was crushed into finer material by the use of metal balls in water.
Bal A word used since medieval times for a group of tin bounds. Thought to be related to the Cornish word *pal*, or a shovel, hence a 'digging'. By the 18th century usually referred to a mine.
Bal maid Term used for female mine surface workers who were engaged in copper and tin dressing.
Baryte or **Barite** Barium sulphate.
Beam work An east Cornwall word for an openwork, where the lode was worked from surface as a trench. *See* Coffin.
Binder *See* Timberman.
Blende *See* Sphalerite.
Blowing-house A building containing a tin-smelting furnace. From at the least the 13th century Cornish blowing-houses had a water-wheel which powered the bellows used to increase the heat in the furnace. From the end of the 17th century they were gradually replaced by reverberatory furnaces.
Borer Old name for drill steel.
Boring machine A 19th-century name for a compressed air-powered rock drill.
Bound or **Tin bound** The area in which tinners worked. Its boundaries or bounds were marked by piles of turfs which were renewed every year when the bounds were re-registered at an itinerant stannary court. Tin bounds almost always had names and the tinners' names were also registered upon payment of a small fee.
Side bounds As the tin bounds were fixed by straight lines between the corners, it was sometimes necessary to register 'side bounds' to include valuable areas outside those lines.
Bowl furnace Early furnaces of simple construction for smelting tin and other metals. The furnace consisted of a pit dug in the ground covered by a domed, stone-built structure. It was lined with clay, had a small *tuyère* or opening to receive a draft and an opening in the top to allow smoke and fumes to escape. The tin concentrate (cassiterite, or black tin) was placed in the furnace with charcoal and heated to a high temperature by means of a draft or hand-operated bellows.
Brass An alloy of copper and zinc.
Bronze An alloy of copper and tin.
Buddle A structure which uses water to separate cassiterite from lighter gangue material. Originally a sloping rectangular box-like structure, but by the middle of the 19th century these were superseded by round buddles that were sometimes convex and sometimes concave. The later 'round frames' were types of buddles.
Burning-house Construction with a furnace for burning off unwanted arsenides and sulphides. Used from at least the 16th century, particularly as more lode material was being processed.
Cage Modern conveyance for carrying men up and down a shaft. Originally referred to the drum carrying the rope of the hoisting gear.
Calamine An ore of zinc. It was used with copper to produce brass, and its presence in the Mendips close to the early 18th-century Bristol brass foundries helped to establish the brass industry there.
Calciner Furnace for roasting ore to remove arsenides and sulphides. William Brunton's and other 19th-century calciners were also designed to remove arsenic to be sold.
Call In the cost-book system of mine management it was frequently necessary to ask adventurers for money to keep the mine going. The amount asked for depended on the number of shares the adventurer owned. The demand for this money was known as a 'call'.
Capstan Winch for hoisting heavy gear up and down the shaft.
Cassiterite Also called tinstone. The

commonest tin ore. Chemical composition tin oxide.

Chalcocite A copper ore. It is grey, metallic and crystalline. Chemical composition copper–iron sulphide.

Chalcopyrite The most common copper ore found in Cornwall. Brassy yellow in appearance. Chemical composition copper–iron sulphide.

Chute or **Shute** Timber or steel structure for facilitating the loading of wagons underground. Usually erected in short 'box-hole' raises beneath stopes. 'Cousin Jacks' was the name given to a widely used design of chute as colonials saw them as typically Cornish.

Clack Non-return valve in the rising main of a typical Cornish pitwork. So called due to the sound it made when closing as the pump rod paused and the weight of water closed the valve.

Coffin, Koffen or **Coghan** Old Cornish word for an openwork, where the lode has been worked from surface as a trench. *See* Beam work.

Coin, Coinage *See* Coinage towns.

Coinage towns Each stannary district in Cornwall had a coinage town where the blocks of white tin were taken to be assayed. A corner ('coin') was cut off each block of tin metal for assaying and the appropriate mark placed on the block. The towns were Helston, Truro, Bodmin and Liskeard, with Lostwithiel being the centre for the whole tin industry. Penzance became a coinage town in the 17th century.

Concentrate Name given to ore after it had been crushed and dressed prior to smelting.

Consols In Cornwall mines with the term 'Consolidated' in the title were invariably known as Consols, e.g. Fowey Consolidated Mines was known as Fowey Consols. Used on its own Consols almost always referred to Consolidated Mines in Gwennap, one of the greatest copper mines in Cornwall.

Convocation *See* Tinners' Convocation or Parliament.

Core Old name for mine shifts – day shift, night shift or afternoon shift.

Cornish roller crushers These were introduced at the beginning of the 19th century to crush copper ore. They were operated either by water power or steam and reduced the ore by crushing it between two steel cylindrical rollers. John Taylor is credited with their introduction.

Cost book company A mining partnership in transferable shares without a limit on the number of shareholders or partners. All expenses and income were entered in the cost book and every month or two months the purser presented the accounts to the adventurers. The adventurers then decided the profits to be shared, the calls to be made and the future policy of the company. An adventurer could give up his share by paying what he owed, or he could sell his shares to whom he wished.

Crasing mill or **Crazing mill** A mill which ground the gravel-size tin ore to a sand or slime consistency using grinding wheels, as in a corn mill. Until the middle of the 16th century they were used after dry stamping, but once wet stamping was introduced and produced a finer pulp, the crasing mill was used at the end of the process for regrinding roasted material or regrinding the tin slag from the blowing-houses.

Croust or **Crowst** The Cornish dialect word for the mid-shift break for food and drink.

Crosscourse Geological fault which cuts across (often at a right-angle) the main direction of lodes. It can vary from a few inches to several hundred yards in width, and can contain valuable minerals. The Great Crosscourse lies beneath the Red River, which divides the parishes of Camborne and Illogan. It was also the boundary between Cooks Kitchen and Dolcoath.

Crosscut Tunnel driven at a right-angle to the lodes in order to gain access to lodes or discover new lodes.

Costean pit Pits dug by tinners in their search for a lode outcrop. Sometimes trenches were used. When a pit exposed a lode it was deepened in order to open up the lode for working. *See* Assay hatch.

Development In mining this refers to the sinking of shafts and winzes and the driving of tunnels on lode or as crosscuts to open up new parts of a mine for production.

Dial Early instrument for mine surveying. The forerunner for the theodolite, it was usually used in conjunction with a compass. By the early 19th century dials carried a level for measuring vertical angles.

Diamond drill Compressed-air drill used for exploration. The bit is studded with industrial diamonds, so that the cut rock passes back through the hollow drill steel and can be recovered and assayed.

Dole Can refer to dues paid to the mineral lord. Also known as the 'lord's dish'. The dole or dues were an agreed proportion of the dressed ore to be paid to the owner of the minerals. Dole can also refer to a share in a tin work or mine.

Dress, Dressing The process by which raw ore is prepared by crushing, sorting or washing into a concentrate ready for smelting. Now called mineral processing.

Dressing floor Cobbled floor on the mine for sorting and dressing the stamped or crushed ore.

Nowadays called the mill. On copper mines these floors could cover very large areas where hundreds of bal maids and youths sorted the copper ore out for assaying. Sometimes the floors had a rough covering or roof to keep the ore and workers dry.

Drill steel Steel bar used to drill rock. Originally bronze and iron were used, but by the 13th century the mine blacksmiths were 'steeling' the cutting ends of the iron borers. By the 1950s most Cornish mine drill steel bits were tipped with tungsten carbide.

Drive or **Drift** Tunnel driven along the course of a lode. 'Drifter' was the name given by Cornish miners to the large, heavy drilling machines introduced in the late 19th century for mine development.

Dry The miner's changing house.

Dues The proportion of the value of ore or metal to be paid to the mineral owner or lord. *See* Dole.

Duty Unit of measurement introduced over 200 years ago to determine a steam engine's performance. It measured the amount of water in pounds lifted one foot by a bushel of coal. Unfortunately, the bushel weight varied considerably, as did the quality of the coal and the honesty of the men doing the calculation.

Eluvial Ore lying on or near the surface downslope from the lode outcrop. Not alluvium. *See* Shoad.

Engine In the late 17th and early 18th century the term 'water engine' came into use. These were pumping engines powered by water. Newcomen's atmospheric engines were called 'fire engines', and this name persisted well into the 19th century. From the 1770s Watt's true steam engines were used, and by the early 19th century the high-pressure Cornish beam engine dominated, as it did until the 20th century.

Engine shaft The shaft which had the mine pumping engine on it, as opposed to shafts which had whim (winding) engines on them.

Fathom English measure of six feet. This was the standard measure both vertically and horizontally in Cornish mines until the second half of the 20th century.

Footwall Wall rock on the lower side of an inclined lode or fault.

Fork Old term for a drained mine. To 'be in fork' was to be fully drained.

Froth flotation In mineral processing, a method of separating out specific material by using its tendency to adhere to froth bubbles. The froth is then skimmed off and the mineral recovered.

Frue vanner A continuous moving inclined belt by means of which the crushed ore is concentrated by a controlled flow of water.

Fullers earth Montmorillonite. A complex clay material.

Galena Lead sulphide. The principal lead and silver ore mined in Cornwall.

Gangue Waste material with no value.

Gig Conveyance for carrying men up and down a shaft.

Gossan The lode at or near surface. Characterised by a dirty red or yellow appearance due to the presence of iron and other minerals.

Grade Measured value of ore after crushing and concentrating. Also refers to value of ore when underground.

Granite Medium- or coarse-grained igneous rock which has intruded into the overlying country rock. Cornish granite and the adjacent country rock can contain valuable metallic mineral lodes or veins.

Grass Cornish miners' name for surface. 'Grass captain' was one in charge of dressing at surface.

Grist mill Corn mill. Technology similar to that used in crazing mills.

Ground In mining can refer to all soil, overburden or rock, whether mineralised or not.

Gunnis An underground void which has been mined out. Sometimes used for a surface openwork or coffin.

Gwyth or **Gweyth** Cornish word for a working. Usually applied to a stream or mine working. In Blackmoor and Foweymore stannaries it was used instead of wheal, which means the same.

Hangingwall Wall rock on the upper side of an inclined lode or fault.

Hatch A small shaft for accessing deep alluvial tin. Can also refer to an exploration or costean pit.

Headgear or **Headframe** Framework over shaft to carry the sheave wheels.

Heading Can refer to the end of a drive or crosscut.

Kaolin Decomposed granite, which has left soft clay and sand.

Kibble Bullet-shaped mine bucket for hoisting ore and waste.

Killas Cornish term for sedimentary rock. The country rock in most of Cornwall.

Knocking mill Another term for a stamping mill. Used more in central and east Cornwall.

Ladderway, Manway or **Man road** Series of ladders which join underground levels or workings and surface.

Lander Man who tends cages or skips at the mouth of a shaft.

Launder A wooden construction for carrying water at surface or underground.

Leat A man-made watercourse. They can be

several miles long and have been used since ancient times in Cornwall. In some western mines the miners used the word for underground gutters.

Level Horizon underground at which tunnels are driven. The 260-fm level refers to all tunnels driven at that horizon. In most Cornish mines levels were given in fathoms, and by the 19th century most levels were spaced at 10, 12 or 20 fathom intervals. These levels were usually measured from below the mine adit level. In most mines the measurement was made down the dip of the lode the main engine shaft was sunk on, hence they were rarely equivalent to a true vertical depth.

Lode Mineralised structure or ore vein.

Lost lovan The trench running from the tail or mouth of the adit portal to the nearest stream.

Man engine Device for carrying men to and from the surface.

Mineral lord Owner of mineral rights, originally held by of lord of the manor. Not the same as land ownership.

Mines Royal Company incorporated by the crown in 1564 to search for various metals in England.

Mispickel *See* Arsenopyrite.

Molybdenite Molybdenum sulphide. The main molybdenum ore mineral.

Mundic Arsenopyrite or mispickel.

Ochre Coloured earthy materials, which are largely oxides and hydrated oxides. A high iron content gives its distinctive yellow, red and brown colouring.

Ore pass Shaft between levels where ore can be sent to the haulage level below.

Overburden Weathered rock and valueless ground which lies above the payable ore.

Pare Team of men led by either a tributer or tutworker.

Pewter An attractive and easily worked metal, which is an alloy of tin (85–95 per cent) and lead. Sometimes other metals like copper are used with the tin. From Roman times until the early modern period pewter ware was the mainstay of the tin industry.

Pillar Ore left to support wall rock in a stope or other excavation.

Pitch, usually **Tribute pitch** Measured part of a stope to be worked by a tributer and his pare.

Pitchblende An important uranium ore consisting mostly of uranium oxide.

Pitman Skilled miner responsible for the pitwork.

Pitwork Pumps and associated gear in the shaft.

Plug-and-feather Pre-gunpowder method of breaking ground. A hole was drilled at an angle into the rock face and two 'feathers', or semi-cylindrical lengths of tapered steel, were inserted into the hole. A metal 'plug' was then hammered into the space between the two feathers, which forced them apart, breaking away the rock to the weaker free face.

Plunger pump Force pump where weight of descending timber rod forces water up a rising main.

Pol pick A miners' pick. One end is sharpened and the other is cut off or polled to form a flat end for use as a hammer. This design has been used in mines since the Bronze Age.

Pre-emption The system by which the crown and then the Duchy of Cornwall had the right to purchase all white tin in the stannaries. They also had the right to farm pre-emption out to foreign bankers or wealthy merchants. The system caused constant controversy and sometimes worked for the good of the tinners and sometimes to their harm.

Raise or **Rise** Relatively narrow passage mined upward. They can connect levels to improve ventilation or be driven into stopes for access or as an ore pass.

Pyrite An iron sulphide mineral.

Reverberatory furnace Type of mineral smelting furnace introduced at the end of the 17th century. Its principal characteristic was keeping the ore separate from the fuel, unlike in a blast furnace.

Rock drill Compressed air-powered machine for drilling shotholes.

Rod When fuses were made by inserting short lengths of goose quill into each other and pouring fine gunpowder into them, they were known as rods. The name persisted until the end of the 20th century for detonators, 160 years after the goose-quill rods had been replaced.

Ruffbudlers Name used in the early 17th century for coarse ore dressers. Primitive buddles were used to wash the ore-bearing material of gangue.

Scoffe or **Skoff mill** Name used in the 17th century for tin mill. Appears either to refer to stamping or crazing mills.

Sett Area of ground leased for mining.

Shaft and level mining System which has been used since ancient times to work lode mines. Vertical or sub-vertical shafts were sunk, usually on the lode, and then levels were driven between them to facilitate ease of working and ventilation.

Shaking table A mechanised ore-dressing apparatus for separating mineralised sand or slime from waste material. Its flat top can be set at a variety of angles and its shaking or vibrating motion can be adjusted in a variety of ways to

suit the metallic ore being separated. Water flows across the table in controlled volume and velocity to suit the material.
Shammel or **Shamble** To move water or rock to surface in stages. Early miners shovelled ore up from bench to bench or stage to stage. Some eighteenth-century pumps lifted water to adit or surface in stages. Steam engines might lift water to a higher level where less powerful water engines could lift it to adit level.
Sheave wheel A large wheel in the headgear, which the hoisting rope passes over, between the drum and the shaft.
Shoad Shoad stone is lode material which has been washed from the outcrop of a lode and lies on the surface down slope from the place it originated. Such tin-bearing materials are also called eluvial deposits.
Shot-hole A hole into which explosives are placed prior to blasting. Originally holes were about 20 in (50 cm) deep and an inch or so in diameter; eventually they could be many metres deep and 3-4 in (7.5-10 cm) in diameter.
Shovell money Late medieval payment made to stannary courts for each registered tinner.
Skip Conveyance for hoisting ore or waste, usually guided by timber or steel runners.
Slag The residual impurities which remained after smelting. These 'zinders' or cinders were usually ground fine for re-dressing.
Smelting The process by which the dressed concentrate is turned into metal. A furnace is used to bring the concentrate to the temperature needed to drive off unwanted material and produce the metal.
Sollar Timber platform in shaft ladderway, a timber platform over a winze or shaft or a false floor in a level to facilitate ventilation. From the Cornish for 'floor'.
Spale A fine. If a miner or tinner was absent or broke the rules he would be spaled or fined. Although used from medieval times, the word was still used at South Crofty in the 1980s, when such fines were usually paid to the local hospital matron.
Spalier, Spallier or **Spalliard** A day labourer. (*See* Spale.) A man employed in a mine or tin stream on a daily basis, usually to replace a tinner or miner who was unable to perform his obligatory work. On tin bounds every shareholder was obliged to work the same number of days. If this was not possible the missing shareholder could send a spalier in his place.
Sphalerite The commonest zinc ore mined in Cornwall. It is a zinc sulphide mineral and was known as blende, Jack, Black Jack or wild lead by Cornish miners.
Stamp or **Stamping mill** Machine for crushing ore. Originally powered by water-wheel but later mostly by powerful steam engines. First mention of water-powered stamps was in Wendron in 1493. Californian and cushion stamps were introduced at the end of the 19th century and powered by electricity. Californian stamps had a cam which turned the head at each descent, enabling the steel heads to last longer. Cushion stamps were faster than the traditional stamps, and an air-cushion prevented them from wear and damage as they stamped the ore.
Stannary district There were four Cornish stannaries, which were administrative districts for controlling and organising the tinners in their areas. The stannary districts were Penwith & Kerrier, Tywarnhaile, Blackmoor and Foweymore. Each of the stannary districts had its own court and there was a great court for the whole of the Cornish stannaries.
Stanniferous Tin-bearing.
Steward The steward of a stannary district was in overall charge of his stannary and was answerable to the Vice-Warden of the Cornish stannaries.
Stope Place from which ore is extracted. Stopes usually lie between the levels and can be worked from below the level (underhand stopes) and from above the level (back stopes). The latter are often called shrinkage stopes because the miner works off the broken ore, which is then pulled through chutes into wagons on the level beneath.
Streaming, Stream works *See* Tin streaming.
Strike The strike of a lode is the direction in which it runs.
Stull Large timbers placed in the workings to support the hangingwall and to form platforms upon which waste rock could be thrown.
Sump The bottom of the pumping or engine shaft.
Sumpman Shaft sinker.
Tail of adit Portal or mouth of an adit where the water runs out at surface.
Tailings Gangue or waste material left after dressing.
Thousand weight Weight used in medieval Cornish tin industry which was 1200 lb (544 kg).
Ticketing System introduced in the 1720s to ensure that all ore was sold. At the ticketing, smelting company's agents would give sealed bids for parcels of copper ore they were interested in. The highest bidder took the parcel, which could be over 100 tons of ore. Collusion between agents meant loss to the mines and the system was occasionally regarded suspiciously.
Timberman or **Binder** Skilled miner

responsible for most safety work underground. They also built chutes, put in ladderways and did some timberwork in shafts. Until the 18th century they were usually called 'binders'. The chief binder was very highly paid.

Tinner Under stannary law the tinner was a privileged worker. He was protected from most feudal restrictions and impositions and could work for tin almost anywhere in unenclosed land, or in enclosed land if it had previously been bounded for tin working. His liberties were protected by royal charters from the time of King John.

Tinners' Convocation or **Parliament** The Charter of Pardon of 1508 gave the tinners the right to hold their own parliament or convocation. Few were held, and those which were were dominated by the wealthy landowners, mine owners and merchants involved in the tin industry.

Tin, Black Black tin is tin concentrate (cassiterite) of usually between 50 and 70 per cent tin metal.

Tin bound *See* Bound.

Tin streaming Producing tin concentrate (cassiterite) from alluvium by using stream water to wash away the lighter gangue material, largely made up of gravel and sand. Until the middle of the 15th century most Cornish and Devon tin came from this source. Name also applied to producing cassiterite from mine tailings, and the hundreds of tin streams in the 19th century were mostly concerned with this rather than true alluvium.

Tin, White White tin is tin metal.

Tinstone Raw tin ore.

Toller An official below the bailiff in the medieval stannary system. In the 18th and 19th centuries he was an official working for mineral lords in collecting dues and registering tin bounds.

Trammer Miner who transported ore and waste to the shaft or ore pass. For much of the 19th century and up until the 1940s trammers pushed wagons of half a ton and 16 cwt along the levels, but from about 1940 electric locos were increasingly used to haul trains of wagons.

Tributer Tributers took a pitch at a price agreed with the mine captain at a Dutch auction, where the lowest bidder usually got the contract. He was paid so much in the pound for the ore he raised based on the current copper standard or price of tin. The tributer normally had to pay all the costs incurred by his work, including paying for powder, fuse, picks, shovels, candles, drill steel sharpening, hoisting up the shaft, sorting and assaying. He also had to pay his pare of men. He was responsible for safety and his men obeying the rules of the mine. He could end up rich or he could earn nothing, but generally he was better off than his fellows.

Tutworker The tutworker was a pieceworker, which is what the old West Country word 'tut' means. He was paid at so much a fathom for the length of tunnel driven or, if he was stoping, by the measured ground he broke. His contract would be re-set at the end of a month or two-month period. The system remained in place in a modified form until the mines closed in 1998.

Tye A word with several meanings in mining and streaming. Can refer to an adit tunnel or a leat at surface. Still in use at South Crofty until the 1980s, when a parallel tunnel or a by-pass tunnel was referred to as a 'side tye'.

Ventilation This was a problem from the moment miners first worked below surface. The ancients solved it mostly by sinking shafts in pairs and driving tunnels between. Many devices or methods were employed over the centuries to cause air to circulate, including putting in false floors (*see* Sollar) and by using the flow of water to move the air. By the 20th century compressed-air and electric fans were used to blow fresher air in to the stopes and ends or draw the hot air out.

Vice-Warden He was in overall charge of the running of the Cornish stannary districts. All difficult cases were brought before him and he presided over the Great Court for the whole of Cornwall. The position was usually held by an important member of the Cornish gentry with an interest in the tin industry. He was answerable to the Lord Warden, who was usually a great magnate and courtier who rarely ventured into Cornwall.

Whim Mine hoist or winder. These could be powered by men, horses, water, steam or electricity. Whim is usually pronounced 'whem'.

Whim plat Circular platform where the horse whim hoist was situated and around which the horse would walk. At the centre was the drum or cage around which the rope was wound. Situated adjacent to the shaft mouth.

Winchester measure or **standard** The government of Wessex introduced standard measurements for length and weight. These standards remained the same for 1000 years until metrication replaced them in the late 20th century. The standard examples were kept at Winchester, the Wessex capital, hence the name.

Winze Excavation driven downwards usually to the next level, either for exploration, ventilation or to form an ore pass.

Wolfram or **Wolframite** Ore of tungsten. Dark brown, almost black, lustrous crystalline mineral. Chemical composition iron–manganese tungstate.

Notes and References

Chapter One The ancient tin industry

1 J.E. Healy, *Mining and Metallurgy in the Greek and Roman World* (1978), p. 70.

2 B. Earl and J.A. Buckley, 'Preliminary report on tin and iron working at Crift Farm', *Journal of the Trevithick Society* (1990), pp. 66–77; J.A. Buckley, M.Phil. thesis, 'Economic significance and role of adits in Cornish mine drainage' (1992), p. 9.

3 R.D. Penhallurick, *Tin in Antiquity* (1986), pp. 173–221.

4 R. Symons, 'Alluvium in Par Valley', *Journal of the Royal Institution of Cornwall* (1877), pp. 382–4; W. Pryce, *Mineralogia Cornubiensis* (1778), p. 68; H. and T. Miles, 'Excavations at Trethurgy, St Austell: interim report', *Cornish Archaeology* (1973), p. 25–9; H. Miles, 'Barrows on the St Austell granite, Cornwall', *Cornish Archaeology* (1975), pp. 5–81.

5 Penhallurick (1986), p. 200; C.F.C. Hawkes and M.A. Smith, 'On some buckets and cauldrons of the Bronze and Early Iron Age', *Antiquities Journal* (1957), pp. 131–98.

6 J.X.W.P. Corcoran, 'Tankard and tankard handles in the British Early Iron Age', *Proceedings of the Prehistoric Society* (1952), pp. 85–102; Penhallurick (1986), p. 221.

7 C.H. Oldfather, *Diodorus of Sicily* (1939), pp. 142, 255; B. Cunliffe, *Pytheas the Greek* (2001), pp. 73–115; J.A. Buckley, *The Cornish Mining Industry* (1992), p. 3.

8 A.C. Thomas, 'The character and origins of Roman Dumnonia', in *Rural Settlements in Roman Britain* (1966), pp. 76–98.

9 Penhallurick (1986), p. 140.

10 Penhallurick (1986), p. 141; G. De Beer, 'Itkin', *Geographical Journal* (1960); Pliny, *Natural History*, Book 4, Ch. 16 (c. 77 ad); E. Green, *Cornwall and the Early Tin Trade* (1917); I.S. Maxwell, 'The location of Ictis', *Journal of the Royal Institution of Cornwall* (1972), pp. 293–319.

11 Penhallurick (1986), pp. 219–21; S. Gerrard, *The Early British Tin Industry* (2000), pp. 15–23.

12 P.B. Ellis, *Caesar's Invasion of Britain* (1978), p. 57; S. Frere, *Britannia* (1973), p. 45; Gerrard (2000), p. 22; J.A. Buckley, 'Historical evidence of alluvial tin streaming in the river valleys of Camborne, Illogan and Redruth', *Journal of the Trevithick Society* (1999), pp. 88–99.

13 B.H. St J. O'Neil, 'Roman villa at Magor, near Camborne', *Devon and Cornwall Notes and Queries* (1932), pp. 40–2. *Acta Sanctorum*, in Penhallurick (1986), p. 245.

14 T. Beare, *The Bailiff of Blackmoor* (1586), edited by J.A. Buckley (1994), p. 1; R. Carew, *Survey of Cornwall*, edited by F.E. Halliday (1969), p. 89; Harleian Manuscript 6380, British Library; J. Hatcher, *English Tin Production and Trade Before 1550* (1973), p. 17.

15 J.A. Buckley, *Medieval Cornish Stannary Charters* (2001).

16 M.H. Stocker, 'Account of some remains found in Pentuan streamworks', *Penzance Natural History and Antiquarian Society* (1952), pp. 88–90; Penhallurick (1986), p. 166; J.D. Muhly, Ph.D. thesis, 'Copper and tin: distribution of mineral resources and the nature of the metal trade in the Bronze Age' (1969), pp. 472–3; Gerrard (2000), p. 21; Buckley, *Cornish Mining Industry* (1992), pp. 3, 4.

17 Penhallurick (1986), pp. 167–9.

18 Gerrard (2000), pp. 19, 20; Earl and Buckley (1990), pp. 66–77.

19 J.A. Buckley, 'Tinners were a race apart', *Western Morning News*, 7 October 1997.

Chapter Two Privilege and rebellion

1 Buckley (2001); G.R. Lewis, *The Stannaries* (1965), p. 34; J. Hatcher, *English Tin Production and Trade Before 1550* (1973), p. 18.

2 T. Clarke, *The Domesday Book* (1996), p. 56.

3 Buckley (2001); Lewis (1965), p. 35.

4 Buckley (2001).

5 Hatcher (1973), p. 24; Lewis (1965), p. 35.

6 Lewis (1965), pp. 16, 35, 43, 96; Buckley (2001).

7 Buckley (2001), pp. 3–6.

8 Buckley (2001), pp. 5–6.

9 *Monasticon Dioecisis Croniensis* no. viii (1170) and no. xix (1250/1); Cornwall Feet of Fines no. 138 (13 October 1251); P.L. Hull, *The Cartulary of St Michael's Mount (1230–40)* (1962).

10 Buckley (2001), pp. 6–9; Lewis (1965), p. 37.

11 Hatcher (1973), pp. 22–6.

12 Buckley (2001), pp. 3–9; Lewis (1965), pp. 38, 39, 109; Hatcher (1973), p. 22; Beare (1586, ed. Buckley 1994), p. 1.

13 Penhallurick (1986), pp. 125, 126; Beare (1586, ed. Buckley 1994), p. 1.

14 Hatcher (1973), p. 22; Buckley (2001), p. 9; N.J.G. Pounds, 'The Duchy Palace at Lostwithiel, Cornwall', *Archaeological Journal* vol. 136 (1979), pp. 203–17; PRO E36/57 42v; PRO E36/37 25r.

15 Beare (1586, ed. Buckley 1994), pp. xiv–xv; Additional manuscript 24746, British Library; Lewis (1965), p. 126.

16 Beare (1586, ed. Buckley 1994), pp. xiv–xv, 3.

17 Beare (1586, ed. Buckley 1994), pp. xiv–xv; Lewis (1965), p. 143.

18 PRO E101/260/4 f3; P. Claughton, 'Silver–lead: a restricted resource: technological choice in Devon silver mines', *Mining Before Powder Peak District Mines Historical Society* (1994), p. 55.

19 Buckley, *Cornish Mining Industry* (1992), pp. 4, 5; PRO E101/260, 261.

20 PRO E101 260/30; E101 260/19 (1306).

21 Buckley (2001), pp. 6–9; Beare (1586, ed. Buckley 1994), pp. 57–60; Lewis (1965), p. 189; Black Prince's Register: White Book of Cornwall (25 Edward III).

22 Lewis (1965), p. 97; Buckley (2001), pp. 12, 13; Hatcher (1973), pp. 62, 63.

23 Hatcher (1973), pp. 59–62.

24 J.A. Buckley, *Tudor Tinbounds* (1987), pp. 10–14; J.A. Buckley, 'Who were the tinners?', *Journal of the Trevithick Society* (1997), pp. 96–105; Buckley (1999), pp. 92–4.

25 Camborne Churchwarden's Accounts (1540s), transcribed by Mrs Brenda Hull; Hatcher (1973), pp. 61, 62; Black Prince's Register: White Book of Cornwall, pp. 109–11; J.A. Buckley *The Great County Adit* (2000), p. 21; Star Chamber Case Henry VIII 125 (Henderson 3/221, RIC).

26 C. Henderson and H. Coates, *Old Cornish Bridges* (1972), pp. 15, 104; Black Prince's Register folio 21 (2 February 1352), folio 34 (10 July 1353), folio 70 (1357), folio 71, folio 78 (1357), folio 98 (1359), folio 100 (1359), folio 111 (1361).

27 C.C. James, *History of Gwennap* (c. 1944), pp. 60, 61, 190.

28 Lewis (1965), p. 252; Hatcher (1973), pp. 23, 91–95; Black Prince's Register: White Book of Cornwall folio 100 (1359); Buckley (2001), pp. 11, 14.

29 Buckley (1987), pp. 9–14, 20; Buckley, 'Who were the tinners?' (1997); Lewis (1965), pp. 195, 196; PRO E101 263; Duchy DCO Rolls Series box 4/481 (1493), Duchy Office, London.

30 Buckley (1987), p. 16; A.L. Rowse, 'The turbulent career of Sir Henry de Bodrugan', *History* vol. xxix (1944), pp. 17–26.

31 Buckley (1987), pp. 9–14; Additional manuscript 24746, British Library; Buckley, 'Who were the tinners?' (1997).

32 Buckley (1987), pp. 10–14; W.J. Blake, *The Cornish Rebellions of 1497, Journal of the Royal Institution of Cornwall* vol. 20, part 1 (1915), pp. 49–86; D. Hay, *Polydore Vergil's Anglica Historia 1485–1537* (1950).

33 Buckley (2001), pp. 15–21; Lewis (1965), pp. 97, 98.

Chapter Three Enterprise in the stannaries

1 Buckley (2000), pp. 18–20; Carew (1602, ed. Halliday 1969), p. 91; Additional manuscript 24746, British Library; D.C. Coleman, *The Economy of England 1450–1750* (1982), pp. 24–6; R.R. Pennington, *Stannary Law* (1973), pp. 11–21.

2 A.L. Rowse, *Tudor Cornwall* (1969), p. 226; Buckley (1987), pp. 10–14; Hatcher (1973), pp. 83, 86; S P Domestic. Elizabeth Addenda xxix 126; A.K.H. Jenkin, *The Cornish Miner* (1948), pp. 124, 125.

3 T.L. Stoate, *The Cornwall Military Survey 1522 with Loan Books and Tinners' Muster of c.1535* (1987); J. Norden, *Description of Cornwall* (1728, ed. Graham 1966), pp. 60, 66; J. Mattingly, 'A tinner and bal maiden: further research on St Neot windows', *Journal of the Royal Institution of Cornwall*

(2001), pp. 96–9.

4 Beare (1586, ed. Buckley 1994), pp. 5, 6, 17, 18, 48; Hatcher (1973), pp. 53, 54; Lewis (1965), pp. 213, 214; Pennington (1973), pp. 224, 225.

5 Rowse (1969), pp. 253–90; J. Chynoweth, *Tudor Cornwall* (2002), pp. 214–26; Lewis (1965), p. 253; A. Fletcher, *Tudor Rebellions* (1983), pp. 40–54; P. Caraman, *The Western Rising 1549* (1994).

6 Beare (1586, ed. Buckley 1994), pp. 6, 37, 56, 57, 59, 60.

7 Beare (1586, ed. Buckley 1994), pp. 60–4.

8 Beare (1586, ed. Buckley 1994), pp. 29, 58–60, 88.

9 Carew (1602, ed. Halliday 1969), pp. 91–5, 228, 229; Norden (1728, ed. Graham 1966), p. 33.

10 Carew (1602, ed. Halliday 1969), pp. 91–3.

11 Carew (1602, ed. Halliday 1969), p. 93.

12 Carew (1602, ed. Halliday 1969), p. 96.

13 Carew (1602, ed. Halliday 1969), pp. 94, 95; Norden (1728, ed. Graham 1966), pp. 13, 14.

14 Lewis (1965), p. 254; Jenkin (1948), pp. 123–6.

15 Jenkin (1948), pp. 126, 127; Lewis (1965), p. 217.

16 G. Agricola (1556), *De Re Metallica*, edited by H. Hoover (1950), pp. 172–99; Buckley, M.Phil. thesis (1992), pp. 54–6.

17 Rowse (1969), pp. 256, 260; Chynoweth (2002), p. 65.

18 C. Hill, *Intellectual Origins of the English Revolution* (1965), p. 16; M.B. Donald, *Elizabethan Copper* (1994), p. 304.

Chapter Four Crown, Parliament and the Glorious Revolution

1 Additional manuscript 24746 p. 121, British Library.

2 W.J. Lewis, *Lead Mining in Wales* (1967), pp. 4–46; Alphabetical Index of Patentees of Inventions (1969), p. 587 (Patent no. 67, 14 January 1634); F.J. Stephens, Typescript at RIC (1927); J. Whetter, *Cornwall in the Seventeenth Century* (1974), p. 69; *Philosophical Transactions of the Royal Society* (1671), p. 2109.

3 Lewis (1965), pp. 218, 219; Whetter (1974), pp. 60, 61; Jenkin (1948), pp. 126, 127.

4 Lewis (1965), pp. 220, 221; *Aggravii Venetiani* (Tinners' Grievance), (1697).

5 N.J.G. Pounds, *The Parliamentary Survey of the Duchy of Cornwall 1649–50* (1982), vol. 1 pp. xiii–xxii, 12–22, 43–53; Coleman (1982).

6 Lewis (1965), pp. 221, 256.

7 Index of Patentees (1969); AR 10/43, CRO (1686); Whetter (1974), p. 23.

8 Pryce (1778), p. 307; Whetter (1974), p. 64; HA/16/17, RIC; William Doidge Tehidy Estate maps, CRO (1737); Additional manuscript 24746, British Library; AR 18/8, CRO (1695).

9 *Philosophical Transactions of the Royal Society* (1671), pp. 2096–113.

10 Agricola (1556/1950); Donald (1967), pp. 300–42; B. Earl, *Cornish Explosives* (1978), pp. 6–20; Carew (1602, ed. Halliday 1969), p. 95; N.J.G. Pounds, 'William Carnsew of Bokelly and his diary 1576–77', *Journal of the Royal Institution of Cornwall* (1978), pp. 14–60; D. Brading and H. Cross, 'Colonial silver mining: Mexico and Peru', *Hispanic American Historical Review* (1972); P.H. Sawyer, *From Roman Britain to Norman England,* (1978), p. 233; Buckley (2000), pp. 27, 28; Breage Parish Burial Registers (1689).

11 C/49/17, RIC; Tehidy Manor (Basset), 'Adventurers and tinbounds' (1739–67), CRO; HA/16/10, RIC; Charles Henderson Index vol. 2, p. 286, RIC; Lewis (1965), pp. 128, 129, 221, 222.

12 C. Morris, *The Journeys of Celia Fiennes* (1947), p. 258; Whetter (1974), pp. 66, 68, 69; CSP Domestic (1680–81), p. 400.

13 Coleman (1982); Buckley, M.Phil. thesis (1992), pp. 49–61.

14 HP/7/1, RIC; Coleman (1982); Buckley, M.Phil. thesis (1992).

15 HA/16/17, RIC.

16 Lease of Goneva alias Wall Worke, Gwinear (August 1699), Private collection.

17 Buckley, *Cornish Mining Industry* (1992), pp. 13, 14.

Chapter Five Copper: birth to boom

1 Norden (1728, ed. Graham 1966), pp. 16, 30; Gerrard (2000), p. 15.

2 Sir John Pettus, *Fodinae Regales* (1670), p. 16; Henry VI Rot.15 and 20 (Sept 10 30 Hen.VI).

3 Donald (1994), pp. 300–42; D.B. Barton, *The History of Copper Mining in Cornwall and Devon* (1978), p. 9.

4 Donald (1994), pp. 316–18, 343–68.

5 Donald (1994), p. 312; *Philosophical Transactions of the Royal Society* (1671), pp. 2096–113.

6 Pryce (1778), pp. 286, 287; J. Houghton, *Collection for Improvement of Trade* (1697), part 2, p. 187; Coleman (1982), pp. 156–8; R. Burt, *The British Lead Mining Industry* (1984), pp. 26, 27; J.B. Richardson, Metal Mining (1974), pp. 13–16.

7 Pryce (1778), pp. 277–9, 286, 287;

8 Barton, *Copper Mining in Cornwall and Devon* (1978), pp. 12–15; W.H. Pascoe, *The History of the Cornish Copper Company* (1981), pp. 20–7.

9 Pryce (1778), p. 287; Barton, *Copper Mining in Cornwall and Devon* (1978), pp. 15–17; Pascoe (1981), p. 24.

10 Pascoe (1981), pp. 21–4; Barton, *Copper Mining in Cornwall and Devon* (1978), p. 15; Pryce (1778), p. 277; R.R. Angerstein, *Travels in England and Wales* (1754), edited by P. Berg (1994), p. 113.

11 J. Brooke, 'Henric Kalmeter's Account of Mining and Smelting in the South-West 1724–1725', (1997), M.Phil. thesis, pp. 289–300; W. Borlase, *Natural History of Cornwall* (1758), p. 206; Pryce (1778), p. 287.

12 DD TEM 60/61, CRO (Carnkye Bal 1771–80); DD TEM 55 (Cooks Kitchen 1781–89); DDX 34, CRO; Tehidy Manor (Basset), 'Adventurers and Tinbounds', CRO (New Dudnance 1753–57), p. 102; DD AD 87, CRO (Wh Gorland 1794–96); DDX 316/2, CRO (Park Friglas and Wh Dudnas 1741–42); DDJ 1788, CRO (Penhellick Worke 1712–19); DDJ 1785, CRO (Pool Adit 1731–37); DD AD 87, CRO; DD TL 94, CRO (Poldice 1789, 1797–98).

13 DD TEM 55, CRO.

14 DD TEM 55, CRO; DDX 475/2-7, CRO (Dolcoath 1771–81).

15 DD TEM 55, CRO; DDX 475/2-7, CRO; DDJ 1788, CRO; DD AD 87, CRO; DDJ 1785, CRO.

16 Pryce (1778), p. 187; Kalmeter, in Brooke (1997), p. 291; South Crofty lease of North Tincroft Sett, 1862.

17 AR/18/8, CRO (1695); Borlase (1758), p. 168, plate xviii; Pryce (1778), p. 172; DDJ 1788, CRO; DDJ 1785, CRO.

18 Pryce (1778), pp. 151, 152, 307, 312, 313; D.B. Barton, *Essays on Cornish Mining History*, vol. 1 (1968), p. 159; A.K.H. Jenkin, *Mines and Miners of Cornwall*, vol. 10 (1965), p. 7.

19 J. Kanefsky, 'The diffusion of power technology in British industry 1760–1870' (1979), Ph.D. thesis; Buckley, M.Phil. thesis (1992), pp. 72–6; *Chronological History of Technology* (1950), p. 6; Barton, *Copper Mining in Cornwall and Devon* (1978), p. 21; D.B. Barton, *The Cornish Beam Engine* (1969), pp. 15–27; Act of Parliament (1741), 'Drawback of the Duties upon Coal for Fire Engines for Draining Tin and Copper Mines in Cornwall'.

20 DDJ 1788, CRO; Jenkin, vol. 10 (1965), pp. 14, 15; Buckley (2000), pp. 32–5.

21 Buckley (2000), pp. 21–4, 35–41; Act of Parliament (1741); Report on Copper Trade (1799); Pigot and Co. New Commercial Directory (1823–24).

22 Buckley (2000).

23 Information supplied by Dr Sharron P. Schwartz.

24 T.R. Harris, 'Some lesser known Cornish engineers', *Journal of the Trevithick Society* vol. 5 (1977), pp. 27–65; Index of Patentees (1969); Barton (1969), pp. 15–27, 84, 137.

25 Pryce (1778), p. xiv, 137; DDX 457/2, CRO.

26 Pascoe (1981), pp. 25, 31; Barton (1969), p. 137; DDJ 1785, CRO; Jenkin (1948), p. 200; Dolcoath Pay Book 1786–91 (20 October 1786), Private collection.

27 Borlase (1758), p. 206.

28 Angerstein (1754/1994), pp. 110–23; Pryce (1778), pp. xi–xiv.

Chapter Six Syndication and amalgamation

1 Barton, *Copper Mining in Cornwall and Devon* (1978), pp. 26, 35–8.

2 Barton, *Copper Mining in Cornwall and Devon* (1978), pp. 38, 39.

3 Barton (1969), pp. 20–7; Barton, *Copper Mining in Cornwall and Devon* (1978), pp. 27–44; Pryce (1778), pp. 308–13.

4 Bryan Earl, pers. comm.

5 Buckley (2000), p. 49; William Jenkin, Letter Books vol. 3, p. 32, and vol. 6, pp. 16, 17, RIC; Barton (1969), pp. 22, 23; Beauchamp Papers (16 August 1784 and 15 July 1791), Private collection; Barton, *Copper Mining in Cornwall and Devon* (1978), pp. 28–38; K.H. Rogers, *The Newcomen Engine in the West of England* (1976).

6 Barton, *Copper Mining in Cornwall and Devon* (1978), pp. 31, 37, 41.

7 *Sherborne Mercury*, 28 February 1791; Barton (1969), pp. 22–7; Barton, *Copper Mining in Cornwall and Devon* (1978), pp. 34, 40.

8 P. Watts-Russell, 'The Cornish Metal Company 1785–92', Paper in preparation.

9 Watts-Russell, op. cit.

10 Watts-Russell, op cit.; Barton, *Copper Mining in Cornwall and Devon* (1978), pp. 36–9.

11 Watts-Russell, op. cit.

12 J.A. Buckley, *Dolcoath*, Publication in preparation; Dolcoath Pay Books DDX 475/2-7; Jenkin, *Mines and Miners of Cornwall*, vols 3, 6, 10; Buckley (2000), pp. 47–9; Report on Copper Trade (1799).

13 A. Raistrick (ed.), *The Hatchett Diary 1796* (1967), pp. 37–41.

14 Jenkin (1948), pp. 152, 298.

15 Pryce (1778), pp. x, xi; Lewis (1965), pp. 256–8; T.R. Harris, *Dolcoath: Queen of Cornish Mines* (1974), p. 24. STA 353 CRO.

16 P. Watts-Russell, 'Cutting the connection' in *Cornish Banner*, May 2003, vol.112; Tin Coinage Resolution Books, 1780–1800.

17 H. Carter, *The Autobiography of a Cornish Smuggler* (1894/1971); Jenkin (1948), pp. 149, 150; G. Borlase, 'Lanisly Letters 1750–56', *Journal of the Royal Institution of Cornwall* (1881), pp. xxiii, 374–9.

18 Jenkin (1948), pp. 149–53.

19 Jenkin (1948), pp. 160, 161.

20 Jenkin (1948), pp. 162–4.

21 Jenkin (1948), p. 154; Tehidy Manor accounts, 12 July 1780 and 28 April 1791, CRO.

Chapter Seven Cornish copper back on top

1 Buckley, *Cornish Mining Industry* (1992), p. 24.

2 Harris (1974), pp. 27–35; Barton, *Copper Mining* (1978), pp. 55–57; J. Lewis, *Fowey Consols: A Richly Yielding Piece of Ground* (1997).

3 Barton, *Copper Mining* (1978), p. 54; Report on Copper Trade (1799); Pigot and Co. New Commercial Directory (1823–24); R. Burt, *John Taylor* (1977), pp. 21–38; Buckley (2000), p. 103.

4 Lewis (1997).

5 J. Lean, *Engine Reporter*, 1811; J. and T. Lean, *Engine Reporters*, 1812–37; B. Howard, *Mr Lean and the Engine Reporters* (2002); T. Lean, *On the Steam Engines in Cornwall* (1839); Harris (1977), pp. 37, 38, 45, 46.

6 Barton (1969), pp. 28–58, 137–68.

7 Lean (1839); Howard (2002).

8 Barton, *Copper Mining* (1978), pp. 55, 56.

9 *West Briton*, 27 September 1822; Day and Night Book of Dolcoath Mine Captains (1822); D.B. Barton, *Tin Mining and Smelting in Cornwall* (1967), pp. 17–71; *Mining Journal,* 17 December 1859, p. 877.

10 Barton (1969), pp. 207–18.

11 Buckley, *Cornish Mining Industry* (1992), pp. 26–9.

12 Buckley, *Cornish Mining Industry* (1992), pp. 26–9.

13 Buckley, *Cornish Mining Industry* (1992), p. 29; C. Noall, *Levant* (1970), pp. 107–17.

14 Buckley, *Cornish Mining Industry* (1992), pp. 29–31; Barton, *Copper Mining* (1978), pp. 47–52.

15 Barton, *Copper Mining* (1978), p. 52; D.B. Barton, *The Redruth and Chasewater Railway 1824–1915* (1978).

16 Barton, *Copper Mining* (1978), p. 53; Buckley, *Cornish Mining Industry* (1992), p. 31.

17 Earl (1978), pp. 6–20; Buckley, *Cornish Mining Industry* (1992), pp. 32, 33; J.A. Buckley, 'The introduction of blasting into Cornish mines', *Wheals Magazine* no. 12 (1982), p. 4.

18 Earl (1978), pp. 6–20; Buckley (1982), p. 4; Breage Parish Burial Registers 1689, 1691.

19 Buckley, *Cornish Mining Industry* (1992), pp. 32, 33.

20 Earl (1978).

21 R. Burt, P. Waite and R. Burnley, Cornish Mines (1987); J.H. Collins, *Observations on the West of England Mining Region* (1912); Barton, *Copper Mining* (1978), pp. 77–96; T. Spargo, *The Mines of Cornwall*, vols 1–6 (1865).

22 Spargo (1865), vol. 5, pp. 14, 16, 17, and vol. 4, pp. 5, 6; Collins (1912).

23 Spargo (1865), vol. 2, pp. 3, 17, 18.

24 DDJ 1788, CRO; DDJ 1784, CRO; DDJ 1785, CRO; J.Y. Watson, *A Compendium of British Mining* (1843), p. 3; J. Rowe, *Cornwall in the Age of the Industrial Revolution* (1993), p. 8; Sir Charles Lemon, 'Statistics of the Copper mines of Cornwall', in R. Burt, *Cornish Mining* (1969), p. 57; J.R.S. Leifchild, *Cornwall, Its Mines and Miners* (1855), p. 173; A.C. Todd, *The Cornish Miner in America* (1967), p. 16; J. Williams, *Cornwall and Devon Mining Directory* (1862); Children's Employment Commissioners (1842); Barton (1967), p. 240.

25 PRO 101 260/30 (34 Edward I 1306).

26 Buckley (1980), pp. 50–2.

27 Day and Night Book of Dolcoath Mine Captains (1822–23), Private collection.

28 Barton, *Copper Mining* (1978), p. 94.

Chapter Eight Change and crisis

1 Barton, *Copper Mining* (1978), pp. 56, 94.

2 Mine reports: South Wheal Frances, North and South Basset, CRO; Burt, Waite and Burnley (1987), pp. 26–9, 191–6; M. Palmer and P. Neaverson, *The Basset Mines: Their History and Industrial Archaeology* (1987).

3 Burt, Waite and Burnley (1987), pp. 78–85, 331, 332.

4 Noall (1970); C. Noall, *Botallack* (1972).

5 Collins (1912), pp. 262–6, 297, 468.

6 Tin Coinage Resolution Books, 1780–1818; Barton (1967), pp. 26–30.

7 Barton (1967), pp. 50, 53, 55, 63, 65, 90, 126, 135; *West Briton*, 26 August 1823.

8 S.P. Schwartz, 'The making of a myth: Cornish miners in the New World in the early 19th Century', *Cornish Studies* no. 9, pp. 105–26; S.P. Schwartz, 'Cornish migration studies', *Cornish Studies* no. 10, pp. 136–65; Dr Sharron P. Schwartz, pers. comm.

9 Dr Sharron P. Schwartz, pers. comm.; Todd (1967); J. Rowe, *The Hard Rock Men: Cornish Immigrants and the North American Mining Frontier* (1973).

10 J. Faull, *Cornish Heritage: A Miner's Story* (1979); P. Payton, *The Cornish Miner in Australia* (1984).

11 S.P. Schwartz, 'Cornish migration to Latin America: a global perspective' (2003), Ph.D. thesis; R. Dawe, *Cornish Pioneers in South Africa: Gold, Diamonds and Blood* (1998).

12 Burt (1977).

13 Buckley (1980); H. Thomas, *Mining Interviews* (1896), pp. 192–204.

14 C. Carter, 'Introduction of compressed air powered rock drills into the Camborne mines', *Journal of the Trevithick Society* (1993), pp. 2–22.

15 Earl (1978).

16 Professor Charles Thomas, pers. comm.

17 J.H. Trounson, *Report on Dolcoath Mine* (1972); Harris (1974).

18 Barton (1967), pp. 136–213.

19 South Wheal Crofty Cost Book, 1891.

20 Barton (1967), pp. 148, 149; *Royal Cornwall Gazette*, 13 January 1872 and 27 January 1872.

21 D. Mudd, *Down Along Camborne and Redruth* (1978), pp. 83, 84.

22 T.A. Morrison, *Cornwall's Central Mines: The Southern District* (1983), p. 40; Harris (1974), pp. 62, 63; Barton (1967), pp. 185, 186.

23 Morrison (1983), pp. 40, 41; Barton (1967), p. 186.

24 Harris (1974), pp. 75–81; Trounson (1972); Buckley, *Cornish Mining Industry* (1992), p. 41; Barton (1967), pp. 195, 218; Mine reports: Dolcoath, Carn Brea and Tincroft, East Pool & Agar, New Cooks Kitchen and Basset Mines, CRO.

Chapter Nine The end of an era

1 *Royal Cornwall Gazette*, 1 January 1870; *Royal Cornwall Gazette*, 21 February 1895; Barton (1967), p. 217.

2 Buckley (1980), pp. 102–19; Harris (1974), pp. 75–83; Morrison (1983), p. 267; Trounson (1972); C. Noall, *Geevor* (1983), p. 80.

3 Buckley, *Cornish Mining Industry* (1992), pp. 41, 42; Barton (1967), pp. 214–49.

4 Tony Brooks, pers. comm.; Buckley (1980), pp. 122, 123.

5 Barton (1967), pp. 235–57; Mine reports: Wheal Peevor, Killifreth, Carn Brea and Tincroft, CRO.

6 Barton (1967), pp. 257–63; Buckley (1980), pp. 144–6; Mine report: South Crofty, South Crofty Archive, Wheal Jane Mine, Truro; Buckley (2000), pp. 85–7.

7 Buckley, *Cornish Mining Industry* (1992), pp. 42–4; Mine reports: Basset Mines, Dolcoath, Wheal Grenville, Geevor, South Crofty and East Pool & Agar, CRO.

8 P. Heffer, *East Pool & Agar: A Cornish Mining Legend* (1985), pp. 36–49.

9 Barton (1967), p. 269; Buckley (1980), pp. 146, 147.

10 Barton (1967), pp. 269–77; Mine reports: Geevor, East Pool & Agar and South Crofty, CRO; Buckley (1980), pp. 146–8.

11 Heffer (1985), pp. 38–49; Barton (1967), p. 269.

12 Buckley, *Cornish Mining Industry* (1992), pp. 42–4; J.A. Buckley, 'A history of Killifreth mine' (1987), unpublished report for Carrick District Council; Barton (1967), pp. 275–7.

13 Buckley (1980), pp. 152–5; Noall (1983), p. 160.

14 Noall (1983), p. 160; Buckley (1980), pp. 156–68; J.A. Buckley, *A Miner's Tale: The Story of Howard Mankee* (1988), pp. 49–53.

15 Buckley (1980), p. 170; Noall (1983), pp. 107–9.

16 Heffer (1885), p. 57; Noall (1983), pp. 108–10; Buckley (1980), pp. 171–6; DDX 161, CRO.

17 Buckley (2000), p. 88; Buckley (1980), pp. 176, 177.

18 Barton (1967), p. 274; Cornish Chamber of Mines Report (1984); Noall (1983), pp. 134–8; J.A. Buckley, *Geevor Mine* (1989).

19 Buckley (1989); J.A. Buckley, South Crofty Mine (1997), pp. 185, 186.

20 Buckley (1989).

21 Buckley, *South Crofty Mine* (1997), pp. 188, 189.

22 Buckley, *South Crofty Mine* (1997), p. 190.

23 Buckley, *South Crofty Mine* (1997), pp. 190, 191.

24 Buckley, *South Crofty Mine* (1997), pp. 191, 192.

Chapter Ten A treasure house of minerals

1 N.G. Leboutillier, *South Crofty: Geology and Mineralisation* (1996); Burt, Waite and Burnley (1987), p. ix.

2 Penhallurick (1986), p. 163; Beare (1586, ed. Buckley 1994), pp. 101–3.

3 Penhallurick (1986), pp. 160–4; H.G. Dines, *The Metalliferous Mining Region of South West England* (1956), vol. 1, pp. 30, 31; Carew (1602, ed. Halliday 1969), p. 88; G. Henwood, *Four Lectures on Geology and Mining* (1855), pp. 8, 9.

4 Dines (1956) vol. 1, pp. 27, 28; Burt, Waite and Burnley (1987), pp. xxviii–xxx, xxxviii, xlvi, 110–12, 303, 413, 418, 429, 499–501; Harris (1974), p. 41; *Transactions of the Royal Geological Society of Cornwall* vol. 6, pp. 112, 113; Noall (1972), p. 119.

5 Dines (1956) vol. 1, pp. 26, 27; Burt, Waite and Burnley (1987), pp. xxvii, xxviii, xxxviii, xlv, 110–12, 303, 413, 419, 420, 499–501; H.L. Douch, *East Wheal Rose* (1979), pp. 42–9; *The Cornish Banner*, 1846.

6 Dines (1956) vol. 1, p. 26; Burt, Waite and Burnley (1987), pp. xxx, xxxviii, xlvii, 56, 111, 112, 344, 413; J.A. Buckley, K.T. Riekstins

and P.R. Deakin, *Wheal Jane Underground* (1997), p. 3.

7 Dines (1956) vol. 1, pp. 28, 29; Burt, Waite and Burnley (1987), pp. xxx–xxxii, xxxviii, il, 172–4, 411, 412, 430, 432.

8 Dines (1956) vol. 1, p. 29; Burt, Waite and Burnley (1987), pp. xxx, xxxi, xxxviii, xlviii.

9 Dines (1956) vol. 1, p. 28; Burt, Waite and Burnley (1987), pp. xxxii, 38, 346, 489.

10 Dines (1956) vol. 1, pp. 23, 24; Burt, Waite and Burnley (1987), pp. xxv–xxvii, xxxviii, xliii, 67, 132, 147, 150; J.A. Buckley, *A History of South Crofty Mine* (1980), pp. 133–77.

11 Dines (1956) vol. 1, pp. 24, 25.

12 Dines (1956) vol. 1, pp. 25, 26; Burt, Waite and Burnley (1987), pp. xxxii, l; Buckley (1980), pp. 122, 128, 129, 134, 190, 193, 198.

13 Dines (1956) vol. 1, pp. 29, 30; Burt, Waite and Burnley (1987), pp. xxxii, xxxiii, 329, 468, 509, 543, 544; C. Noall, *The St Ives Mining District*, vol. 2 (1993), pp. 65–8.

14 Dines (1956) vol. 1, p. 30; Burt, Waite and Burnley (1987), p. xxxii; Thyssen Review (Crofty Consultancy archive).

15 Dines (1956) vol. 1, p. 30; Burt, Waite and Burnley (1987), p. xxxii; Thyssen Review.

16 Dines (1956) vol. 1, p. 30; Burt, Waite and Burnley (1987), p. xxxiii; Thyssen Review.

17 Dines (1956) vol. 1, p. 30; Burt, Waite and Burnley (1987), p. xxxiii; Thyssen Review.

18 Dines (1956) vol. 1, pp. 31, 32; Burt, Waite and Burnley (1987), pp. xxxiii; Thyssen Review.

Chapter Eleven China clay and slate

1 R.M. Barton, *A History of the China-Clay Industry* (1966); J.R. Smith, *Cornwall's China-Clay Heritage* (1992); C. Thurlow, *China Clay from Cornwall and Devon* (2001); C. Thurlow, *China Clay* (1996); Ivor Bowditch, pers. comm.

2 C. Vincent, *Delabole Slate Quarry: Past and Present* (1984); Karl Forfar, pers. comm.

3 *Western Morning News*, 1 December 1952.

4 Carew (1602, ed. Halliday 1969), p. 87.

5 Capt. J. Jenkin, *The History of Old Delabole Slate Quarry* (late 19th century).

6 *Western Morning News*, 1 December 1952.

7 Anon, 'The Delabole 'Ole', Newspaper article, April 1987.

8 *Western Morning News*, 1 December 1952.

Chapter Twelve Cornish miners and mining into the future

I take great pleasure in acknowledging the information, help and advice I have received from all those involved with the several companies referred to or described in this chapter. I am grateful to Peter Sheppard (Drillserve), Charlie and Robin Daniel (South Western Mining and Tunnelling), Mike Hallewell (Holman-Wilfley), Derek Morgan, Dale Foster, Jem Williamson, Karla Riekstins (Crofty Consultancy), Roger Wedlake (Carnon Contracting), Clifford Rice (Wheal Jane Laboratory), Jonathan Hore (Saxton), Alan Pope, John Laity (Seacore), Kevin Williams, David Stone and Alan Reynolds (Baseresult). These men are justly proud of what their companies have achieved since the closure of the mines in Cornwall, and they are all working hard to secure a future for the skills acquired in the mines, quarries and foundries of the Duchy.

Bibliography

Books

Anon (1950), *Chronological History of Technology* London (Science Museum)

Agricola, G. (1556), *De Re Metallica*, edited by H. Hoover (1950) New York (Dover)

Angerstein, R.R. (1754), *Travels in England and Wales*, edited by P. Berg (1994) London (Science Museum)

Barton, D.B. (1967), *Tin Mining and Smelting in Cornwall* Truro (Bradford Barton)

Barton, D.B. (1968), *Essays on Cornish Mining History*, vol. 1. Truro (Bradford Barton)

Barton, D.B. (1969), *The Cornish Beam Engine* Truro (Bradford Barton)

Barton, D.B. (1978), *The Redruth and Chasewater Railway 1824–1915* Truro (Bradford Barton)

Barton, D.B. (1978), *The History of Copper Mining in Cornwall and Devon* Truro (Bradford Barton)

Barton, R.M. (1966), *A History of the China-Clay Industry* Truro (Bradford Barton)

Beare, T. (1586), *The Bailiff of Blackmoor*, edited by J.A. Buckley (1994) Camborne (Penhellick)

Borlase, W. (1758), *Natural History of Cornwall* Oxford (W Jackson)

Buckley, J.A. (1980), *A History of South Crofty Mine* Redruth (Truran)

Buckley, J.A. (1987), *Tudor Tinbounds* Redruth (Truran)

Buckley, J.A. (1988), *A Miner's Tale: The Story of Howard Mankee* Camborne (Penhellick)

Buckley, J.A. (1989), *Geevor Mine* Penzance (Geevor Tin Mines Ltd)

Buckley, J.A. (1992), *The Cornish Mining Industry* Redruth (Tor Mark)

Buckley, J.A. (1997), *South Crofty Mine* Redruth (Truran)

Buckley, J.A. (2000), *The Great County Adit* Camborne (Penhellick)

Buckley, J.A. (2001), *Medieval Cornish Stannary Charters* Camborne (Penhellick)

Buckley, J.A., *Dolcoath*, Publication in preparation.

Buckley, J.A., K.T. Riekstins and P.R. Deakin (1997), *Wheal Jane Underground* Camborne (Penhellick)

Burt, R. (1977), *John Taylor* Buxton (Moorland)

Burt, R. (1984), *The British Lead Mining Industry* Redruth (Truran)

Burt, R., P. Waite and R. Burnley (1987), *Cornish Mine* Exeter (Exeter University Press)

Caraman, P. (1994), *The Western Rising 1549* Tiverton (Westcountry Books)

Carew, R. (1602), *Survey of Cornwall*, edited by F.E. Halliday (1969) New York (Augustus M Kelly)

Carter, H. (1894/1971) *The Autobiography of a Cornish Smuggler* Truro (Bradford Barton)

Chynoweth, J. (2002), *Tudor Cornwall* Stroud (Tempus)

Clarke, T. (1996), *The Domesday Book* Godalming (Coombe Books)

Coleman, D.C. (1982), *The Economy of England 1450–1750* London (Oxford University Press)

Collins, J.H. (1912), *Observations on the West of England Mining Region* Truro (Cornish Mining Classics)

Cunliffe, B. (2001), *Pytheas the Greek* London (Allen Lane)

Dawe, R. (1998), *Cornish Pioneers in South Africa: Gold, Diamonds and Blood* St Austell (Cornish Hillside)

Dines, H.G. (1956), *The Metalliferous Mining Region of South West England*, vol. 1 London (HM Stationery Office)

Donald, M.B. (1994), *Elizabethan Copper* Ulverston, Cumbria (Red Earth Publications)

Douch, H.L. (1979), *East Wheal Rose* Truro (Bradford Barton)

Ellis, P.B. (1978), *Caesar's Invasion of Britain* London (Book Club Associates)

Faull, J. (1979), *Cornish Heritage: A Miner's Story* Adelaide (J Faull)

Fletcher, A. (1983), *Tudor Rebellions* Harlow, Essex (Longman)

Frere, S. (1973), *Britannia* London (Book Club Associates)

Gerrard, S. (2000), *The Early British Tin Industry* Stroud (Tempus)

Green, E. (1917), *Cornwall & the Early Tin Trade* Bath (George Gregory)

Harris, T.R. (1974), *Dolcoath: Queen of Cornish Mines* Camborne (Trevithick Society)

Hatcher, J. (1973), *English Tin Production and Trade Before 1550* Oxford (Clarendon)

Hay, D. (1950), *Polydore Vergil's Anglica Historia 1485–1537* London (Camden Series, Roy.Hist.Soc)

Healy, J.E. (1978), *Mining and Metallurgy in the Greek and Roman World* London

Heffer, P. (1985), *East Pool & Agar: A Cornish Mining Legend* Redruth (Truran)

Henderson, C., and H. Coates (1972), *Old Cornish Bridges* Truro (Bradford Barton)

Hill, C. (1965), *Intellectual Origins of the English Revolution* London (Panther)

Houghton, J. (1697), *Collection for Improvement of Trade* London (J Houghton)

Howard, B. (2002), *Mr Lean and the Engine Reporters* Camborne (Trevithick Society)

Hull, P.L. (ed.) (1962), *The Cartulary of St Michael's Mount* Exeter (Devon & Cornwall Record Office)

James, C.C. (*c.* 1944), *History of Gwennap* Redruth (C C James)

Jenkin, A.K.H. (1948), *The Cornish Miner* London (Allen & Unwin)

Jenkin, A.K.H. (1962–65), *Mines and Miners of Cornwall*, vols 3, 6 & 10 Truro (Truro Book Shop)

Jenkin, Capt. J. (*c.*1890s), *The History of Old Delabole Slate Quarry* Launceston (C H Eveleigh)

Lean, T. (1839), *On the Steam Engines in Cornwall*, reprint (1969) Truro (Bradford Barton)

Leboutillier, N.G. (1996), *South Crofty: Geology and Mineralisation* Camborne (Penhellick)

Leifchild, J.R.S. (1855), *Cornwall, Its Mines and Miners* London (Longman)

Lewis, G.R. (1965), *The Stannarie* Truro (Bradford Barton)

Lewis, J. (1997), *Fowey Consols: A Richly Yielding Piece of Ground* St Austell (Cornish Hillside)

Lewis, W.J. (1967), *Lead Mining in Wales* Cardiff (University of Wales)

Morris, C. (1947), *The Journeys of Celia Fiennes* London (Cresset Press)

Morrison, T.A. (1983), *Cornwall's Central Mines: Southern District* Penzance (Alison Hodge)

Mudd, D. (1978) *Down Along Camborne & Redruth* Bodmin (Bossiney Books)

Noall, C. (1970), *Levant* Truro (Bradford Barton)

Noall, C. (1972) *Botallack* Truro (Bradford Barton)

Noall, C. (1983), *Geevor* Penzance (Geevor Tin Mines Ltd)

Noall, C. (1993), *The St Ives Mining District*, Vol. 2. Redruth (Truran)

Norden, J. (1966), *Description of Cornwall*, edited by Frank Graham, Newcastle Upon Tyne (F Graham)

Oldfather, C.H. (1939), *Diodorus of Sicily* (Loeb Classical Library)

Palmer, M., and P. Neaverson (1987), *The Basset Mines: Their History and Industrial Archaeology* Sheffield (Northern Mines Research Society)

Pascoe, W.H. (1981), *The History of the Cornish Copper Company* Redruth (Truran)

Payton, P. (1984), *The Cornish Miner in Australia* Redruth (Truran)

Penhallurick, R.D. (1986), *Tin in Antiquity* London (Institute of Metals)

Pennington, R.R. (1973), *Stannary Law* Newton Abbot (David & Charles)

Pettus, Sir John (1670), *Fodinae Regales* London (Thomas Basset)

Pounds, N.J.G. (1982), *The Parliamentary Survey of the Duchy of Cornwall 1649–50* Exeter (Devon & Cornwall Record Office)

Pryce, W. (1778), *Mineralogia Cornubiensis* London (William Pryce)

Raistrick, A. (1976), *Charles Hatchett Diary 1796* Truro (Bradford Barton)

Richardson, J.B. (1974), *Metal Mining* London (Allen Lane)

Rogers, K.H. (1976), *The Newcomen Engine in the West of England* Bradford-on-Avon (Moonraker Press)

Rowe, J. (1973), *The Hard Rock Men: Cornish Immigrants & the North American Mining Frontier* Liverpool (Liverpool University Press)

Rowe, J. (1993), *Cornwall in the Age of the Industrial Revolution* St Austell (Cornish Hillside)

Rowse, A.L. (1969), *Tudor Cornwall* New York (Charles Scribner's Sons)

Sawyer, P.H. (1978), *From Roman Britain to Norman England* London (Methuen)

Smith, J.R. (1992), *Cornwall's China-Clay Heritage* Truro (Twelveheads Press)

Spargo, T. (1865), *The Mines of Cornwall* (6 vols.) reprinted (1960) Truro (Bradford Barton)

Stoate, T.L. (1987), *The Cornwall Military Survey 1522 with Loan Books & Tinners' Muster of c.1535* Bristol (T L Stoate)

Thomas, H. (1896), *Cornish Mining Interviews* Camborne (Camborne Printing & Stationery Company)

Thurlow, C. (1996), *China Clay* Redruth (Tor Mark)

Thurlow, C. (2001), *China Clay from Cornwall and Devon* St Austell (Cornish Hillside)

Todd, A.C. (1967), *The Cornish Miner in America* Truro (Bradford Barton)

Trounson, J.H. (1972), *Report on Dolcoath Mine* Camborne (South Crofty Mine Ltd)

Vincent, C. (1984) *Delabole Slate Quarry: Past and Present* Wadebridge (C Vincent)

Watson, J.Y. (1843), *A Compendium of British Mining* London (J Y Watson)

Whetter, J. (1974), *Cornwall in the Seventeenth Century* Padstow (Lodenack)

Williams, J. (1862), *Cornwall & Devon Mining Directory* London (Kent & Co.)

Journal articles, etc.

Blake, W.J. (1915), 'The Cornish rebellions of 1497', *Journal of the Royal Institution of Cornwall* vol. 20, part 1.

Borlase, George (1881), 'Lanisly letters 1750–56', *Journal of the Royal Institution of Cornwall*, pp. 374–9.

Brading, D., and H. Cross (1972), 'Colonial silver mining: Mexico and Peru', *Hispanic American Historical Review*

Brooke, J. (1997), 'Henric Kalmeter's Account of Mining and Smelting in the South-West 1724-1725, M. Phil thesis, Exeter University

Buckley, J.A. (1982), 'The introduction of blasting into Cornish mines', *Wheals Magazine* no. 12, p. 4.

Buckley, J.A. (1992), 'Economic significance and role of adits in Cornish mine drainage'. M.Phil. thesis, Cambourne School of Mines, Exeter University.

Buckley, J.A. (1997), 'Who were the tinners?', *Journal of the Trevithick Society*, pp. 96–105.

Buckley, J.A. (1997), 'Tinners were a race apart', *Western Morning News*, 7 October 1997.

Buckley, J.A. (1999), 'Historical evidence of alluvial tin streaming in the river valleys of Camborne, Illogan and Redruth', *Journal of the Trevithick Society*, pp. 88–99.

Carter, C. (1993), 'Introduction of compressed air powered rock drills into the Camborne mines', *Journal of the Trevithick Society*, pp. 2–22.

Claughton, P. (1994), 'Silver–lead: a restricted resource: technological choice in Devon silver mines', *Mining Before Powder Peak District Mines Historical Society*, p. 55.

Corcoran, J.X.W.P. (1952), 'Tankard and tankard handles in the British Early Iron Age', *Proceedings of the Prehistoric Society*, pp. 85–102.

De Beer, G. (1960), 'Itkin', *Geographical Journal*.

Earl, B., and J.A. Buckley (1990), 'Preliminary report on tin and iron working at Crift Farm', *Journal of the Trevithick Society*, pp. 66–77.

Harris, T.R. (1977), 'Some lesser known Cornish engineers', *Journal of the Trevithick Society* vol. 5, pp. 27–65.

Hawkes, C.F.C., and M.A. Smith (1957), 'On some buckets and cauldrons of the Bronze and Early Iron Age', *Antiquities Journal*, pp. 131–98.

Kanefsky, J. (1979), 'The diffusion of power technology in British industry 1760–1870'. Ph.D. thesis, Exeter University.

Mattingly, J. (2001), 'A tinner and bal maiden: further research on St Neot windows', *Journal of the Royal Institution of Cornwall*, pp. 96–9.

Maxwell, I.S. (1972), 'The location of Ictis', *Journal of the Royal Institution of Cornwall*, pp. 293–319.

Miles, H. (1975), 'Barrows on the St Austell granite, Cornwall', *Cornish Archaeology*.

Miles, H., and Miles, T. (1973), 'Excavations at Trethurgy, St Austell: interim report', *Cornish Archaeology*, pp. 25–9.

Muhly, J.D. (1969),'Copper and tin: distribution of mineral resources and the nature of the metal trade in the Bronze Age'. Ph.D. thesis, Yale University.

O'Neil, B.H. St J. (1932), 'Roman villa at Magor, near Camborne', *Devon and Cornwall Notes and Queries*, pp. 40–2.

Pounds, N.J.G. (1978), 'William Carnsew of Bokelly and his diary 1576–77', *Journal of the Royal Institution of Cornwall*, pp. 14–60.

Pounds, N.J.G. (1979), 'The Duchy Palace at Lostwithiel, Cornwall', *Archaeological Journal* vol. 136, pp. 203–17.

Rowse, A.L. (1944), 'The turbulent career of Sir Henry de Bodrugan', *History* vol. xxix, pp. 17–26.

Stocker, M.H. (1952), 'Account of some remains found in Pentuan streamworks', *Penzance Natural History and Antiquarian Society*, pp. 88–90.

Schwartz, S.P. (2003), 'Cornish migration to Latin America: a global perspective'. Ph.D. thesis, Exeter University.

Schwartz, S.P., 'The making of a myth: Cornish miners in the New World in the early 19th Century', *Cornish Studies* no. 9, pp. 105–26.

Schwartz, S.P., 'Cornish migration studies', *Cornish Studies* no. 10, pp. 136–65.

Symons, R. (1877), 'Alluvium in Par Valley', *Journal of the Royal Institution of Cornwall*, pp. 382–4.

Thomas, A.C. (1966), 'The character and origins of Roman Dumnonia', *Rural Settlements in Roman Britain*, pp. 76–98. CBA Research Report no. 7.

Watts-Russell, P. (2002), 'Cutting the connection', Cornish Banner, May 2003, vol. 112.

Watts-Russell, P., 'The Cornish Metal Company 1785–92', Paper in preparation.

Archival sources

British Library, London.

Cornwall Record Office (CRO), Truro, Cornwall.

Morrab Library, Penzance.

Public Record Office/The National Archives (PRO), Kew, Surrey.

Royal Institution of Cornwall (RIC), Truro, Cornwall.

South Crofty Archive, Truro.

List of Subscribers

The publishers are grateful to the following individuals and institutions for their support in the development of this book.

Acton, Bob and Stephanie
Devoran, Cornwall
Acworth, Ian, Kim and Emma
Melrose, MA, USA
Acworth, Revd. Dr Richard
Havant, Hampshire
Adams, Anne
Dauntsey Lock, Wiltshire
Adey, T F
Paignton, Devon
Allison, Susan
Ireby, Cumbria
Alvarez-Buylla, Mrs Mary
Guildford, Surrey
Ambler, John
Wisborough Green, West Sussex
Andrewartha, Mr Warwick
Caulfield, VIC, Australia
Andrews, James Warren
Golden, CO, USA
Annear, John Marshall
Dulwich, London
Annear, Nicholas Marshall Poon
Dulwich, London
Annear, Mark O'Donovan Poon
Dulwich, London
Argall, Squadron Leader Ian H A
St Keverne, Cornwall
Ashley, Robert W,
Wendouree, VIC, Australia
ASKi UK Ltd
Marazion, Cornwall
Axton, Bryan Edward
West Molesey, Surrey
Bailey, Edward and Trudy
Five Lanes, Cornwall
Baker, Owen A
Rosevidney, Cornwall
Ball, Diana P
Killivose, Cornwall
Banfield, Keith
Mitcham, SA, Australia
Barr, Thomas
Mitcham, Surrey
Barker, Ashley
Penzance, Cornwall
Beardsell, Andrew B
Brighouse, West Yorkshire
Beautyman, Paul
An Agaidh Mhòr, Scotland
Beeman, R J
Polwheveral, Cornwall
Beer, Keith E
Doddiscombsleigh, Devon
Behnke, Dan
Northbrook, IL, USA
Bell, Professor Stephen C
Western Park, Leicester
Bendle, Margaret
Newquay, Cornwall
Bennett, John
Hayle, Cornwall
Berryman, Bill
Stafford, Staffordshire

Besanko, Christopher
Sutton Coldfield, West Midlands
Birrell, Ralph W
Strathfieldsaye, VIC, Australia
Bishop, Mr Andrew
Wimbledon, London
Blackman, Tony
Perranporth, Cornwall
Blake, Frederick David
Redruth, Cornwall
Bluck, Mick
Southern River, WA, Australia
Bolitho, Dr Elaine E
Ngaio, Wellington, New Zealand
Bolitho, Hilary
Ilminster, Somerset
Bolitho, Mrs E M
Tremethick Cross, Cornwall
Borthwick, Don
Gateshead, Tyne and Wear
Botterill, Anne and Deryck
Torquay, Devon
Bowden, Harold H
Callington, Cornwall
Bowden, Colin
Great Hallingbury, Hertfordshire
Bratton, A C
Oxford, Oxfordshire
Bray, Robert
Bradley Stoke, Bristol
Breakspear, Michael
Box, Wiltshire
Brewer, Collin William
Sladesbridge, Cornwall
Brewer, Frank L
Mevagissey, Cornwall
Bridges (nee Paull), Mrs Shelagh J R
Ovington, Northumberland
Bridle ACSM, BSc(Hons), Ralph
Dhahran, Saudi Arabia
Bristow, Colin
Carlyon Bay, Cornwall
Brown, Mrs Rosemary
St Austell, Cornwall
Brown, Wella
Saltash, Cornwall
Bruce, Rob
Enfield, Middlesex
Burgess, D J R
Allet, Cornwall
Burley, Christine
Canterbury, Kent
Burrow, K J
Bideford, Devon
Burrow, Michael J
Rothwell, Leeds
Busby, Graham
St Mellion, Cornwall
Cacheris (nee Warne), Jan
Phoenix, AZ, USA
Cannon, Philip
Marsh, Buckinghamshire
Carbis, In-Pensioner John C
Chelsea, London
Carew Pole, Sir Richard
Torpoint, Cornwall
Carr, Martin
Penzance, Cornwall

Carter, J E
Tweed Heads West, NSW, Australia
Carwithen, F A
Addleston, Surrey
Chantry, P A
St Austell, Cornwall
Chapman, Mr and Mrs Jim
Launceston, Cornwall
Chapman, David
Gunnislake, Cornwall
Child, Veronica
Camborne, Cornwall
Clarke (nee Annear), Rosemary Doreen
Tavistock, Devon
Cleaves, Mr Stephen James
Radstock, Somerset
Clynick, P R R
Crewkerne, Somerset
Cock, Owen W N
Pinner, Middlesex
Cockerham, Paul
Wendron, Cornwall
Cocks, Ainsley
Par, Cornwall
Coode, Alan
St Austell, Cornwall
Cook, Rex
Nelson, Lancashire
Cook, Robert
Longford, TAS, Australia
Coomb, Mrs D E
Portscatho, Cornwall
Coombe, Canon Michael
Exmouth, Devon
Coon, George Vernon
Ashburton, Devon
Copsey, Brian
Bushey Heath, Hertfordshire
Corbet, Sally
St Buryan, Cornwall
Corin, Carol
St Ives, Cornwall
Corn, Andrew
Busselton, WA, Australia
Cornish Studies Library
Redruth, Cornwall
Cornwall County Council, Planning, Transportation and Estates Dept., Truro, Cornwall
Cosby, Rev Ivan P S G
Bodinnick-by-Fowey, Cornwall
Cothey, Dr V J
St Ives, Cornwall
Coutts, John and Madeleine
Orpington, Kent
Cowling, Mark Graham
New Plymouth, Taranaki, New Zealand
Cowling, Russell Maurice
New Plymouth, Taranaki, New Zealand
Creasy, Mark G
Mosman Park, WA, Australia
Croggon, Richard and Janice
Buninyong, VIC, Australia
Crompton, John Thomas
Blagill, Cumbria

Crosfill, Martin
Heamoor, Cornwall
Crossley, Anthony
Greenwich, NSW, Australia
Cullis, Paul J
Folkestone, Kent
Curnow, Howard
St Hilary, Cornwall
Dadda, Miss J A
Poole, Dorset
Dally, Lindsay J
Mt Waverley, VIC, Australia
Darkin, David
Illogan, Cornwall
Davidson, Pauline Yvonne
Salisbury, Wiltshire
Davies, Mrs J
Bude, Cornwall
Davis, Jan
Lemon Grove, CA, USA
Day, Andrew
Bodmin, Cornwall
Deutsches Bergbau-Museum
Bochum, Germany
Dingle, Mrs Mary
Bundaberg, QLD, Australia
Dore, Patricia A
Redding, CA, USA
Dowell ACSM, Mark
Lilli Pilli, NSW, Australia
Downing, C D
Falmouth, Cornwall
Drew, John R
San Jose, CA, USA
Drew FRSA, Robert G
West Tolgus, Cornwall
Dudley, Terry (Member of Trevithick Society)
Newport, South Wales
Dukes, Kenneth
Churchill, Oxon
Dymond, A J
Launceston, Cornwall
Eade, Patricia M
Glen Waverley, VIC, Australia
Eastlake, Les
St Tudy, Cornwall
Ebling, Albert
Downs Barn, Milton Keynes
Eddy FRCS FRCOG, John W
Colchester, Essex
Ede, Roger M
Kingskerswell, Devon
Edmonds, E W A
Carnon Downs, Cornwall
Eich, Clive
Luxulyan, Cornwall
Ellis (Johnson), Jocelyn
The Lizard, Cornwall
Fairhurst, Arthur
Maidenhead, Berkshire
Field, Robert and Ceza
Brightwell cum Sotwell, Oxon
Finnemore, Michael J
Ashby de la Zouch, Leicestershire
Foster, Joyce
Truro, Cornwall
Friday, Nigel
Exmouth, Devon

Frost, Peter R
Jurbise, Belgium
Fryer, Rosalyn and David
Polgooth, Cornwall
Garwood, Paul
Hinckley, Leicestershire
Gaved, Arthur E
Rilla Mill, Cornwall
George, Richard L
North Huntingdon, PA, USA
Gibbs, Mary Payne
Denver, CO, USA
Giles, T D B
Porthpean, Cornwall
Ginn, Douglas
Welling, Kent
Gill, Mike
Sutton in Craven, West Yorkshire
Gillis, Ann
Pencorse, Cornwall
Gould, Malcolm, Sue and Lowenna
St Blazey, Cornwall
Graham, Michael I S
Long Marston, Warwickshire
Gray, James
Norton Sub Hamdon, Somerset
Greeves, Dr Tom
Tavistock, Devon
Grimshaw, Kareen
Te Aroha, North Island, New Zealand
Halsey BEng ACSM, M G
Falmouth, Cornwall
Hancock, Helen
Barry, Vale of Glamorgan
Hancock, M T G
Barry, South Glamorgan
Harris, Michael
Biscovey, Cornwall
Harris, Gavin
Biscovey, Cornwall
Hastings, Colin and Helen
Portscatho, Cornwall
Hawthorn, Vince
Fetcham, Surrey
Heeley, Dr Edward
Wennington, Lancaster
Higgins, John
Curitiba, Parana, Brazil
Hobart, John
Ludgvan, Cornwall
Hockin, J C
Swanage, Dorset
Hodge, James
Penzance, Cornwall
Holland, Sir Geoffrey
St Ives, Cornwall
Holland-Smith, Michael
Feock, Cornwall
Hollest, R H
Shaw, Wiltshire
Holman, H
Perranarworthal, Cornwall
Holmes, Jonathan James
Pendeen, Cornwall
Hoskin, Peter
Mawson, ACT, Australia
Hosking, William Rex
Hawthorn, SA, Australia
Hoskins, John H
Sioux Falls, SD, USA
Hughes, John Vivian
Port Talbot, West Glamorgan
Hughes, Susan M
Tregadillett, Cornwall
Hunter, Ed
Victor, CO, USA
Hutchens, Marshall
Plain-an-Gwarry, Cornwall
Hutcheon, Andrew and Sheila
Ivybridge, Devon
Jackman, Brenda
Penryn, Cornwall
James, John F
London
James, Lincoln R
Falmouth, Cornwall
Jarvis, Ann
Oakley, Bedfordshire
Jee, Eric A
Devoran, Cornwall
Jelbart, Ralph D
Catford, London
Jenkin, Ann Trevenen
Hayle, Cornwall
Jennings, William R
Pasadena, CA, USA
Jewell, C R
Bude, Cornwall
Johns, Mr C
Hayle, Cornwall
Jose, John
Llanfarian, Aberystwyth
Juleff, Kitto
Appledore, Kent
Juleff, Lyn
Kenron, QLD, Australia
Kessell, Bettina Grace
Bathurst, NSW, Australia
Kinsmen, Revd Barry
Padstow, Cornwall
Kitto, Jeff
Stockton, CA, USA
Kitto, Robert J
West Lakes, SA, Australia
Kopp, Rita Bone
Ashland, OH, USA
Lakin, Bill
Pendeen, Cornwall
Lane, Peter and Lesley
Tremar, Cornwall
Lanyon, Mr C J E
St Just in Roseland, Cornwall
Leach FSA, Peter E
St Mawes, Cornwall
Leggat, Dr and Mrs Peter
West Looe, Cornwall
Lerk, James A
Golden Square, VIC, Australia
Libby, David J
Santiago, Chile
Lizut, John
Lipson, Devon
Lloyd, Philip E
Manchester
Lloyd, Leanne
Eaglehawke, VIC, Australia
Lokan (nee Goldsworthy), Dr Jan
McLaren Vale, SA, Australia
Long, Anne C M
St Just, Cornwall
Lorigan, Catherine
Reading, Berkshire
Luff, M J
Ravenstone, Leicestershire
Luke, Tom and Libby
Wantirna, VIC, Australia
MacMillan, N C and S E
Bundanoon, NSW, Australia
Maddern, Allan J
Andover, Hampshire
Mainwaring, Ross
St Ives, NSW, Australia
Makin, K
Walsden, Lancashire
Mallett, Elizabeth
Gladwin, MI, USA
Malyan, Hugh D
Hayle, Cornwall
Manderston-Mackrill, Mrs B
Truro, Cornwall
Marchant, Bob Le
Lower Kelly, Cornwall
Marshall, Brian Roberts
Farnham, Surrey
Marshall, Barbara
Wollongong, NSW, Australia
Martin Jr, James Memory
Brookings, SD, USA
Martin, Kevin and Judy
Fowey, Cornwall
Martyn, Dr Kendal
Cambridge
Matthews, Duncan Paul
Liskeard, Cornwall
Mayers, Lynne
Blaize Bailey, Gloucestershire
McGivern, Adrian
Oakley, Bedfordshire
McGuire White, Penny
Mentone, VIC, Australia
McKay CEng CSci MIMMM MCSM, Dr Angus
Inverurie, Scotland
McIntosh, Linda Simmons
Honolulu, HI, SA
Medlyn, John
Mickleton, Gloucestershire
Meeson, John and Ann
Bodrigan, Cornwall
Michell, Kenneth James
Currabubula, NSW, Australia
Michell, L W
St Mary's, Isles of Scilly
Michell, Donovan
Gorran, Cornwall
Miller, Chris
Castle Cary, Somerset
Miller, Ken
Weymouth, Dorset
Millerchip, John D
Spring, TX, USA
Mitchell, Elizabeth
West Melton, Christchurch, NZ
Mitchell, Rosemary
Cumberland Park, SA, Australia
Moore, Michael
Perth, WA, Australia
Moore, Victor
Evans, GA, USA
Morgan (nee Harvey), Mrs R O
West Wickham, Kent

LIST OF SUBSCRIBERS

Morris, William A
London
Moyle, Terry
South Darenth, Kent
Muhlhaus, Dr John
Great Bookham, Surrey
Mulligan Jr, Dr William H
Murray, KY, USA
Mynott, Robert F
Trumpington, Cambridge
National Museum of Wales
(Library) Cardiff, Wales
Nethercott, Elizabeth
Torpoint, Cornwall
Nethersole, Nigel E
Redruth, Cornwall
Newby, Bill
Lelant, Cornwall
Newman, Phil
Newton Abbot, Devon
Newsham, D F
Billinge, Lancashire
Nicholas, Hanna
Leytonstone, London
Nobbs, Richard
Kingston upon Thames, Surrey
Norman, Marilyn
Port MacDonnell, SA, Australia
Nurhonen, Jon
St Erth Praze, Cornwall
Nuttall, Ralph
Belfast, Northern Ireland
Oke, Graham
Albury, NSW, Australia
O'Rell, Michael
Manhattan Beach, CA, USA
Palamountain, Brian Anthony
Atawhai, Nelson,
New Zealand
Parnell, David and June
Worcester Park, Surrey
Parsons, Derek
Maker, Cornwall
Pascoe, J D
Etobicoke, ON, Canada
Paul, Peter T
Red Cliffs, VIC, Australia
Payn, Mr D S
Mevagissey, Cornwall
Payton, Professor Philip
Bodmin, Cornwall
Peachey, Dominic
Mitcham, Surrey
Pearse, Mike
Mirfield, West Yorkshire
Peerless, Ken
Laity Moor, Cornwall
Pengelly, Jim
Arnold, Nottingham
Penrose, Dominick
Stithians, Cornwall
Pentreath, Dr R J
Bath, Somerset
Perranzabuloe Old Cornwall Society
Perranporth, Cornwall
Perriam, Mike
Buckland in the Moor, Devon
Phillips, Nicky
Hertford, Hertfordshire
Phillips, John A
Kendal, Cumbria
Phillips, Mr Donald E
Callington, Cornwall
Phillips, Wilfred T
Fareham, Hampshire
Pile, James Devereux
Farmington, NM, USA
Piper, Tony
Camborne, Cornwall
Playle, Mr John
Hammersmith, London
Polglase, Stephen
Greatwork, Cornwall
Porter, Philip
Kings Langley, Hertfordshire
Powell, Dr B D
Northfield, Birmingham
Probert, John C C
Redruth, Cornwall
Proffitt, David
Looe, Cornwall
Prowse OBE, Irwin
Page, ACT, Australia
Pullen, Dave
Brackendowns, Alberton,
South Africa
Punchard, S
Tiverton, Devon
Queensland, Cornish Association of
Wishart, QLD, Australia
Raddy, Darren James
West Looe, Cornwall
Raddy, Luke Adam
West Looe, Cornwall
Raddy, Laura-Jane
West Looe, Cornwall
Read, William Edward
Clevedon, North Somerset
Rednall, Richard
Peebles, Scotland
Reynolds, Michael G
Knaphill, Surrey
Richards, Geoff and Angela
Egloshayle, Cornwall
Rickard, James M W
Fanling, NT, Hong Kong
Riggs, Jeffrey Barrington
Plympton, Devon
Riley, M J
Forrest Hill, Auckland,
New Zealand
Roach, Dewey (desc of Capt'n Sam Curnow)
Phoenix, AZ, USA
Roberts, Colin
Trentham, Staffordshire
Robinson, Dorothy J
Avoca, VIC, Australia
Rogers, Allan B
Boulder, CO, USA
Rogers, Mary H
Devizes, Wiltshire
Rogers, Jeremy
Sandhurst, Berkshire
Rovellotti, Fabrizio
Quarona, VC, Italy
Rowe Jr, Harry M
Hemet, CA, USA
Rowe, Lawson R
Burpengary, QLD, Australia
Rule, Laurence
Camborne, Cornwall
Rule, Tony
Helston, Cornwall
Russell, C J G
St Peter Port, Guernsey
Rutter, W
Falmouth, Cornwall
Ryder, D H
Keyham, Devon
Sanders, Judith Gail Rickard
Santa Rosa, CA, USA
Sandoe, Jill
Stockton on Forest, Yorkshire
Schoolar, Ian R
Coventry, Warwickshire
Searle, Peter C
Oakwood, Leeds
Serpell, Nick
London
Shaw, R P
Aylestone, Leicester
Shaw, Richard
Baldock, Hertfordshire
Shephard, Colin
Nancledra, Cornwall
Shipton, Guy
Penzance, Cornwall
Simmons, David J
Pooraka, SA, Australia
Simpson, Dr D A
Burnside, SA, Australia
Sivell, Steve
London
Sleeman, Oliver C
Thirlmere, NSW Australia
Smale, Mr D C
Chelmsford, Essex
Smith, Dr F W
Spennymore, Durham
Smitheram, William H
Santa Barbara, CA, USA
Snedden, Professor Richard J
Malvern, VIC, Australia
South Australian Genealogy and Heraldry Society Inc
Unley, SA, Australia
Sovereign Hill Museums Association,
The Ballarat, VIC, Australia
Sowery, Mr C A
Trewoon, Cornwall
Speed-Andrews, Jonathan
Alton, Hampshire
Speight, Alan and Janet
Yatton, North Somerset
Spencer, Lloyd
Cubert, Cornwall
Spriggs, Professor Matthew
Canberra, ACT, Australia
Standing, Judith
Bateman, WA, Australia
Stephens, W J
Prestwood,
Buckinghamshire
Stephens, Dr F Graham
Portreath, Cornwall
Stephens, Les and Anne
Lake Hopatcong, NJ, USA
Stirk, Carole
Crawley, West Sussex
Stuthridge ACSM FIMM, L A
St Stephen, Cornwall

Sutcliffe, E Mary
Whittlesford, Cambridge
Sutcliffe, Barbara
Nelson, Lancashire
Swaine, Chris
Richmond, VIC, Australia
Swiggs, John Noel
Par, Cornwall
Symons, John C
Malvern Wells, Worcestershire
Taperell, Ken
South Benfleet, Essex
Tarry, Norman
Thurlby, Lincolnshire
Tatam, Ian P
St Minver, Cornwall
Taylor, Stuart
Truro, Cornwall
Terrell, David S
North Saanich, BC, Canada
Thomas, David
Nantwich, Cheshire
Thomas, Joe
Illogan, Cornwall
Thompson, Woodrow B
Wayne, ME, USA
Thompson, Rita
London
Thorne, Graham
Maldon, Essex
Toms, Don
Lead, SD, USA
Tonkin, W John
St Austell, Cornwall
Tonkin, Lindsay and Marc
Eldene, Wiltshire
Tonking, Michael J H
White River, South Africa
Trebilco, Major Peter
Waterloo, NSW, Australia
Tredennick, JoAnn
Jacksonville, FL, USA
Tregoning, Burnett
Mill Valley, CA, USA
Tregoning, J G
Lundy Bay, Cornwall
Treleaven, Mr M A
St Teath, Cornwall
Treloar, Peter Q
Calne, Wiltshire
Trembath, T J
St Just, Cornwall
Trenerry, Walter Northey
W. St Paul, MN, USA
Trengove, Pamela
Wadebridge, Cornwall
Trerise, Bert
Port Angeles, WA, USA
Tresidder, Mark
Cranfield, Bedfordshire
Trevelyan, Raleigh
St Veep, Cornwall
Trevenna, Tim and Val
St Dennis, Cornwall
Treverrow, Barry
Trethurgy, Cornwall
Trevethan, Lt-Cdr F R
Liskeard, Cornwall
Trevithick, Mark E
Littleton, CO, USA
Trevivian, Jacqueline
Veryan, Cornwall
Treweek, Miss S
Sutton, Surrey
Trewhella, John
Marazion, Cornwall
Trezise, Rose and Ashleigh
Hampstead, London
Unger, Elizabeth Buschlen
Manhattan, KS, USA
University of Nevada Reno
The Reno, NV, USA
Varker, John L R
St Austell, Cornwall
Venning, John
Callington, Cornwall
Verso, Jean
Hurstbridge, VIC, Australia
Vigars, Robert
London
Voaden, Dr Denys J
College Park, MD, USA
Wadeson, Tim
Llandudno, South Africa
Wakeford, Robert John
Burcott, Buckinghamshire
Wall, J B
St Austell, Cornwall
Ward, Patrick John
Tolgullow, Cornwall
Warne III, James Ernest
Phoenix, AZ, USA
Warne Jr, James Ernest
Phoenix, AZ, USA
Warren, Kenneth J
Penzance, Cornwall
Warrington, Dr Geoff
Radcliffe on Trent, Nottingham
Warry, P J
Pilton, Somerset
Waters, Malcolm Foster
St Austell, Cornwall
Watkins, Vicky
Kehelland, Cornwall
Watson, Sir Bruce
Fir Tree Pocket, QLD, Australia
Watts-Russell, Penny
Barrow-on-Soar, Leicestershire
Westgate, Benjamin Mark
Mile End, Gloucestershire
Whiffin, June
Blackburn South, VIC, Australia
White, Alfie
Rostrenen, Brittany, France
White, Martin
Exmouth, Devon
White, C J
Ilkley, Yorkshire
Whitford JP MRAeS, Percy
Datchet, Berkshire
Wiblin, C N
St Leven, Cornwall
Wiggin, Geoff
Carbis Bay, Cornwall
Willcocks, D R
St Cleer, Cornwall
Williams, Mr and Mrs Michael C
Probus, Cornwall
Williams, Col. G T G
St Tudy, Cornwall
Williams, Roger
Marazion, Cornwall
Williams, James P
Scorrier, Cornwall
Williams, Derek R
Oswestry, Shropshire
Williams, John (Ace)
Hamlyn Heights, VIC, Australia
Willies, Lynn
Matlock Bath, Derbyshire
Wills, Margaret and Trevor
Gulval, Cornwall
Wilson, Mrs Anne
Waltham Cross, Herfordshire
Wilson, Elizabeth M E
Sunbry, VIC, Australia
Wonnacott ACSM '85, Dr Gavin
Bedford, Bedfordshire
Woolf, David
Richmond, Surrey
Woods, Mary Kinder
Grantham, Lincolnshire
Woon, C D
London
Wright, Iain J
Chichester, West Sussex
Younger, Prof. Paul L
Birtley, Co. Durham
Zuber, Hans and Nicci
Fowey, Cornwall

Index

Note: References in *italic* are to maps and captions to illustrations. References in **bold** are to the Glossary.

Author's Acknowledgements

There are a great many people who have contributed to this book, some knowingly and some unknowingly. I have used the late Roger Penhallurick's book *Tin in Antiquity* extensively for my first chapter. Before its publication we had long discussions on aspects of the archaeological record, and I am truly grateful for his enlightening comments. Charles Smith and Professor Ken Hoskin contributed their unique knowledge of Cornish tin mineralisation to the discussions. John Hatcher's *English Tin Production and Trade before 1550* has also been a very useful guide to the medieval tin trade. *The Stannaries* by G. R. Lewis and *Stannary Law* by R. R. Pennington were used extensively on stannary organisation and practice in the medieval period. The writings of Professor Barry Cunliffe, and the many discussions with Bryan Earle and Professor Ronnie Tylecote have helped me considerably to understand details about those ancient tinners and how they worked. The writing of and conversations with Professor Charles Thomas, Dr Peter Claughton and the late Leslie Douch have also helped me with my researches. The books of Dr John Chynoweth, Dr A. L. Rowse, Professor M. B. Donald, Dr Christopher Hill and Dr A. K. Hamilton Jenkin helped enormously in writing the chapters covering the 16th to 18th centuries, and I am grateful to each of those authors, as we must all be to John Norden and Richard Carew for their wonderful descriptions of Cornish mining over 400 years ago and to William Borlase and William Pryce for their incomparable descriptions of the industry in the 18th century.

I would also like to thank Dr Joanna Mattingly, Dr Bernard Deacon, Dr Sharron Schwartz, Dr Oliver Padel, Professor Roger Burt, Dr Sandy Gerrard, Jack Trounson, Bridget Howard, Penny Watts-Russell, Eric Edmonds, Tony Brooks, Justin Brooke, Clive Carter, Joe Thomas, Michael Tangye, Paul Richards and Alan Reynolds for the help they have given me in my researches, either through their writings or by personnel communication. Among the many miners and technical staff I would like to thank are Peter Hughes, Gerald Pengilly, Kevin Ross, Mike Hodgson, John Polglase, Robin Boon, Charlie Flood, Ronnie Opie, Reggie Moyle, Howard Mankee, Henry Kaczmarek, Freddie Calf, Howard Vigus, Martin Crothers, Roy Thomas, Eric Eckersall, Jeff Sullivan, Cyril Penrose, Jack Jervis, Fred Mounsey, Clarence Matthews, David Eustace, David Ellisdon, Neil Hitchens and dozens of others who have helped me over the years to learn about mining and its history, traditions and techniques. I am grateful for the support that Dale Foster, Karla Riekstins and Derek Morgan have given me as this work progressed, particularly for the time they willingly gave to help me locate maps and use their computers. The list of people who have helped with explanations, arguments or general comments seems endless, and so I hope that those who have been inadvertently missed from those named will forgive me.

To Terry Knight and the staff at the Cornwall Centre, Redruth, Angela Broome and the staff at the Courtney Library, Truro, and Paul Brough and the archivists at the Cornwall Record Office, Truro, I extend my gratitude. They have all always been more than willing to help and go 'that extra mile' to find the material needed or help with an explanation. Nicholas Johnson and his fellow archaeologists at the Cornwall Archaeological Unit (Historic Environment) have also been most helpful. During the last quarter-of-a-century I am sure I have taken up a disproportionately large amount of time from each of those organisations, and I am very grateful to all of them.

To my publishers, Ian Grant and Cornwall Editions, I extend my thanks for giving me the opportunity to indulge my long-held wish to attempt to put this fascinating story, covering perhaps 4000 years, into a single continuous narrative. Lastly, I would like to thank my wife, Sonia, for her continued patience over many months, as I struggled to bring this book to fruition.

Allen Buckley (September 2005)

Picture Credits

Every effort has been made to trace the copyright holders of the illustrations but this has not always been possible. The Publishers would be grateful for any further information on copyright sources of any uncredited illustrations.

Abbreviations: AC ARKA Cartographics; **CRO** Cornwall Record Office; **HES** Historic Environment Service, Cornwall County Council; **JAB** J.A. Buckley; **RIC** Royal Institution of Cornwall, Royal Cornwall Museum.
t = top; b= bottom; l = left; r = right

2, 9 P. R. Deakin; 10-11 AC; 13-19 RIC; 20 HES; 21 RIC; 24-25 John Wotton; 27 David Ashby; 31 Bryan Earle; 32-33 AC; 37 RIC; 38t David Ashby; 38b HES; 41 RIC; 42-43 HES; 49 Allen & Unwin; 50 AC; 51 RIC; 52 Dover Pub; 54 Colonial Williamsburg Foundation; 56-57 Trinity College Cambridge; 61 RIC; 63 RIC; 64-65 Dover Pub; 67 RIC; 72-73 CRO; 74-76 JAB; 76-79 RIC; 81 John Wotton; 82 AC; 83-84 RIC; 88-90 CRO; 93-97 RIC; 101 JAB; 103 CRO; 105 T. Cuneo est.; 106 David Ashby; 108 Trevithick Society; 109-11 CRO; 115-17 RIC; 119 CRO;120 RIC; 122 AC; 123-33 RIC; 134 CRO; 135-42 RIC; 143 AC; 147 RIC; 149 T. Cuneo est.; 150-52 RIC; 154 CRO; 155-68 RIC; 169 T. Cuneo est.; 170 H. G. Ordish; 171 RIC; 175-76 P. R. Deakin; 179-85 RIC; 186 British Geological Society; 189 St Ives Museum; 190-91 RIC; 193 Penlee House Gallery and Museum; 194-95 RIC; 197 Penlee House Gallery and Museum; 198-203 RIC; 207 JAB; 211 John Wotton; 213 P. R. Deakin.